John Allen Paulos

INNUMERACY

Mathematical Illiteracy
and Its Consequences

VIKING

VIKING

Published by the Penguin Group
27 Wrights Lane, London W8 5TZ, England
Viking Penguin Inc., 40 West 23rd Street, New York, New York 10010, USA
Penguin Books Australia Ltd, Ringwood, Victoria, Australia
Penguin Books Canada Ltd, 2801 John Street, Markham, Ontario, Canada L3R 1B4
Penguin Books (NZ) Ltd, 182–190 Wairau Road, Auckland 10, New Zealand

Penguin Books Ltd, Registered Offices: Harmondsworth, Middlesex, England

First published in the USA by Hill and Wang, a division of Farrar, Straus and Giroux,
New York 1988
Published simultaneously in Canada by Collins Publishers, Toronto
First published in Great Britain by Viking 1989
10 9 8 7 6 5 4 3 2

Printed and bound in Great Britain by
Richard Clay Ltd, Bungay, Suffolk

A CIP catalogue record for this book is available from the British Library

ISBN 0-670-83008-9

9017

This book is to be returned on or before the last date stamped below.

19. JAN 1990

1 3 JAN 2009

15. FEB 1990

23 MAR 1990

-1 JUN 1990

-5. JUL 1995

Paulos, J.A.

Contents

Introduction / 3

1 *Examples and Principles* / 6

2 *Probability and Coincidence* / 25

3 *Pseudoscience* / 49

4 *Whence Innumeracy?* / 72

5 *Statistics, Trade-Offs, and Society* / 99

Close / 133

To Sheila, Leah, and Daniel

for numberless reasons

Innumeracy

Introduction

"Math was always my worst subject."

"A million dollars, a billion, a trillion, whatever. It doesn't matter as long as we do something about the problem."

"Jerry and I aren't going to Europe, what with all the terrorists."

. . .

Innumeracy, an inability to deal comfortably with the fundamental notions of number and chance, plagues far too many otherwise knowledgeable citizens. The same people who cringe when words such as "imply" and "infer" are confused react without a trace of embarrassment to even the most egregious of numerical solecisms. I remember once listening to someone at a party drone on about the difference between "continually" and "continuously." Later that evening we were watching the news, and the TV weathercaster announced that there was a 50 percent chance of rain for Saturday and a 50 percent chance for Sunday, and concluded that there was therefore a 100 percent chance of rain that weekend. The remark went right by the self-styled grammarian, and even after I explained the mistake to him, he wasn't nearly as indignant as he would have been had the weathercaster left a dangling participle. In fact, unlike other failings which are hidden, mathematical illiteracy is often flaunted: "I can't even balance my checkbook." "I'm a people person, not a numbers person." Or "I always hated math."

Part of the reason for this perverse pride in mathematical ignorance is that its consequences are not usually as obvious as are those of other weaknesses. Because of this, and because I firmly believe that people respond better to illustrative particulars than they do to general exposition, this book will examine many real-world examples of innumeracy—stock scams, choice of a spouse, newspaper psychics, diet and medical claims, the risk of terrorism, astrology, sports records, elections, sex discrimination, UFOs, insurance and law, psychoanalysis, parapsychology, lotteries, and drug testing among them.

I've tried not to pontificate excessively or to make too many sweeping generalizations about popular culture or our educational system (à la Allan Bloom), but I have made a number of general remarks and observations that I hope are supported by the examples. In my opinion, some of the blocks to dealing comfortably with numbers and probabilities are due to quite natural psychological responses to uncertainty, to coincidence, or to how a problem is framed. Others can be attributed to anxiety, or to romantic misconceptions about the nature and importance of mathematics.

One rarely discussed consequence of innumeracy is its link with belief in pseudoscience, and the interrelationship between the two is here explored. In a society where genetic engineering, laser technology, and microchip circuits are daily adding to our understanding of the world, it's especially sad that a significant portion of our adult population still believes in Tarot cards, channeling mediums, and crystal power.

Even more ominous is the gap between scientists' assessments of various risks and the popular perceptions of those risks, a gap that threatens eventually to lead either to unfounded and crippling anxieties or to impossible and economically paralyzing demands for risk-free guarantees. Politicians are seldom a help in this regard since they deal with public opinion and are therefore loath to clarify the likely hazards and trade-offs associated with almost any policy.

Because the book is largely concerned with various inadequacies—a lack of numerical perspective, an exaggerated appreciation for meaningless coincidence, a credulous acceptance of pseudosciences, an inability to recognize social trade-offs, and so on—much of the writing has a debunking flavor to it. Nevertheless, I hope I've avoided the overly earnest and scolding tone common to many such endeavors.

The approach throughout is gently mathematical, using some elementary ideas from probability and statistics which, though deep in a sense, will require nothing more than common sense and arithmetic. Some of the notions presented are rarely discussed in terms accessible to a wide audience and are the kind of thing that my students, for example, often enjoy but usually respond to with: "Will we need to know that for the quiz?" There won't be a quiz, so they can be enjoyed freely, and the occasional difficult passage can be ignored with impunity.

One contention of the book is that innumerate people characteristically have a strong tendency to personalize—to be misled by their own experiences, or by the media's focus on individuals and drama. From this it doesn't necessarily follow that mathematicians are impersonal or formal. I'm not, and the book isn't either. My goal in writing it has been to appeal to the educated but innumerate—at least to those whose fear of mathematics is not so great that (num)(ber) is automatically read as (numb)(er). The book will have been well worth the effort if it can begin to clarify just how much innumeracy pervades both our private and our public lives.

1

Examples and Principles

Two aristocrats are out horseback riding and one challenges the other to see which can come up with the larger number. The second agrees to the contest, concentrates for a few minutes, and proudly announces, "Three." The proposer of the game is quiet for half an hour, then finally shrugs and concedes defeat.

A summer visitor enters a hardware store in Maine and buys a large number of expensive items. The skeptical, reticent owner doesn't say a word as he adds the bill on the cash register. When he's finished, he points to the total and watches as the man counts out $1,528.47. He then methodically recounts the money once, twice, three times. The visitor finally asks if he's given him the right amount of money, to which the Mainer grudgingly responds, "Just barely."

The mathematician G. H. Hardy was visiting his protégé, the Indian mathematician Ramanujan, in the hospital. To make small talk, he remarked that 1729, the number of the taxi which had brought him, was a rather dull number, to which Ramanujan replied immediately, "No, Hardy! No, Hardy! It is a very interesting number. It is the smallest number expressible as the sum of two cubes in two different ways."

Big Numbers, Small Probabilities

People's facility with numbers ranges from the aristocratic to the Ramanujanian, but it's an unfortunate fact that most are on the aristocrats' side of our old Mainer. I'm always amazed and depressed when I encounter students who have no idea what the population of the United States is, or the approximate distance from coast to coast, or roughly what percentage of the world is Chinese. I sometimes ask them as an exercise to estimate how fast human hair grows in miles per hour, or approximately how many people die on earth each day, or how many cigarettes are smoked annually in this country. Despite some initial reluctance (one student maintained that hair just doesn't grow in miles per hour), they often improve their feeling for numbers dramatically.

Without some appreciation of common large numbers, it's impossible to react with the proper skepticism to terrifying reports that more than a million American kids are kidnapped each year, or with the proper sobriety to a warhead carrying a megaton of explosive power—the equivalent of a million tons (or two billion pounds) of TNT.

And if you don't have some feeling for probabilities, automobile accidents might seem a relatively minor problem of local travel, whereas being killed by terrorists might seem to be a major risk when going overseas. As often observed, however, the 45,000 people killed annually on American roads are approximately equal in number to all American dead in the Vietnam War. On the other hand, the seventeen Americans killed by terrorists in 1985 were among the 28 million of us who traveled abroad that year—that's one chance in 1.6 million of becoming a victim. Compare that with these annual rates in the United States: one chance in 68,000 of choking to death; one chance in 75,000 of dying in a bicycle crash; one chance in 20,000 of drowning; and one chance in only 5,300 of dying in a car crash.

Confronted with these large numbers and with the correspondingly small probabilities associated with them, the innumerate will inevitably respond with the non sequitur, "Yes, but what if you're that one," and then nod knowingly, as if they've demolished your argument with their penetrating insight. This tendency to personalize is, as we'll see, a characteristic of many people who suffer from innumeracy. Equally typical is a tendency to equate the risk from some obscure and exotic malady with the chances of suffering from heart and circulatory disease, from which about 12,000 Americans die each week.

There's a joke I like that's marginally relevant. An old married couple in their nineties contact a divorce lawyer, who pleads with them to stay together. "Why get divorced now after seventy years of marriage? Why not last it out? Why now?" The little old lady finally pipes up in a creaky voice: "We wanted to wait until the children were dead."

A feeling for what quantities or time spans are appropriate in various contexts is essential to getting the joke. Slipping between millions and billions or between billions and trillions should in this sense be equally funny, but it isn't, because we too often lack an intuitive feeling for these numbers. Many educated people have little grasp for these numbers and are even unaware that a million is 1,000,000; a billion is 1,000,000,000; and a trillion, 1,000,000,000,000.

A recent study by Drs. Kronlund and Phillips of the University of Washington showed that most doctors' assessments of the risks of various operations, procedures, and medications (even in their own specialties) were way off the mark, often by several orders of magnitude. I once had a conversation with a doctor who, within approximately twenty minutes, stated that a certain procedure he was contemplating (*a*) had a one-chance-in-a-million risk associated with it; (*b*) was 99 percent safe; and (*c*) usually went quite well. Given the fact that so many doctors seem to believe that there must be at least eleven people in the waiting room if they're to avoid being idle, I'm not surprised at this new evidence of their innumeracy.

For very big or very small numbers, so-called scientific notation is often clearer and easier to work with than standard notation and I'll therefore sometimes use it. There's nothing very tricky about it: 10^N is 1 with N zeroes following it, so 10^4 is 10,000 and 10^9 is a billion. 10^{-N} is 1 divided by 10^N, so 10^{-4} is 1 divided by 10,000 or .0001 and 10^{-2} is one hundredth. 4 × 10^6 is 4 × 1,000,000 or 4,000,000; 5.3 × 10^8 is 5.3 × 100,000,000 or 530,000,000; 2 × 10^{-3} is 2 × 1/1,000 or .002; 3.4 × 10^{-7} is 3.4 × 1/10,000,000 or .00000034.

Why don't news magazines and newspapers make appropriate use of scientific notation in their stories? The notation is not nearly as arcane as many of the topics discussed in these media, and it's considerably more useful than the abortive switch to the metric system about which so many boring articles were written. The expression $7.39842 × 10^{10}$ is more comprehensible and legible than seventy-three billion nine hundred and eighty-four million and two hundred thousand.

Expressed in scientific notation, the answers to the questions posed earlier are: human hair grows at a rate of roughly 10^{-8} miles per hour; approximately $2.5 × 10^5$ people die each day on earth; and approximately $5 × 10^{11}$ cigarettes are smoked each year in the United States. Standard notation for these numbers is: .00000001 miles per hour; about 250,000 people; approximately 500,000,000,000 cigarettes.

Blood, Mountains, and Burgers

In a *Scientific American* column on innumeracy, the computer scientist Douglas Hofstadter cites the case of the Ideal Toy Company, which stated on the package of the original Rubik cube that there were more than three billion possible states the cube could attain. Calculations show that there are more than 4 × 10^{19} possible states, 4 with 19 zeroes after it. What the package says isn't wrong; there are more than three billion possible states. The understatement, however, is symptomatic of a pervasive innumeracy which ill suits a technologically based society. It's analogous to a sign at the entrance to the Lincoln Tunnel stating:

New York, population more than 6; or McDonald's proudly announcing that they've sold more than 120 hamburgers.

The number 4×10^{19} is not exactly commonplace, but numbers like ten thousand, one million, and a trillion are. Examples of collections each having a million elements, a billion elements, and so on, should be at hand for quick comparison. For example, knowing that it takes only about eleven and a half days for a million seconds to tick away, whereas almost thirty-two years are required for a billion seconds to pass, gives one a better grasp of the relative magnitudes of these two common numbers. What about trillions? Modern Homo sapiens is probably less than 10 trillion seconds old; and the subsequent complete disappearance of the Neanderthal version of early Homo sapiens occurred only a trillion or so seconds ago. Agriculture's been here for approximately 300 billion seconds (ten thousand years), writing for about 150 billion seconds, and rock music has been around for only about one billion seconds.

More common sources of such large numbers are the trillion-dollar federal budget and our burgeoning weapons stockpiles. Given a U.S. population of about 250 million people, every billion dollars in the federal budget translates into $4 for every American. Thus, an annual Defense Department budget of almost a third of a trillion dollars amounts to approximately $5,000 per year for a family of four. What have all these expenditures (ours and theirs) bought over the years? The TNT equivalent of all the nuclear weapons in the world amounts to 25,000 megatons, or 50 trillion pounds, or 10,000 pounds for every man, woman, and child on earth. (One pound in a car, incidentally, demolishes the car and kills everyone in it.) The nuclear weapons on board just one of our Trident submarines contain eight times the firepower expended in all of World War II.

To cite some happier illustrations for smaller numbers, the standard I use for the lowly thousand is a section of Veterans Stadium in Philadelphia which I know contains 1,008 seats and which is easy to picture. The north wall of a garage near my house contains almost exactly ten thousand narrow bricks. For one hundred thousand, I generally think of the number of words in a good-sized novel.

To get a handle on big numbers, it's useful to come up with one or two collections such as the above corresponding to each power of ten, up to maybe 13 or 14. The more personal you can make these collections, the better. It's also good practice to estimate whatever quantity piques your curiosity: How many pizzas are consumed each year in the United States? How many words have you spoken in your life? How many different people's names appear in *The New York Times* each year? How many watermelons would fit inside the U.S. Capitol building?

Compute roughly how many acts of sexual intercourse occur each day in the world. Does the number vary much from day to day? Estimate the number of potential human beings, given all the human ova and sperm there have ever been, and you find that the ones who make it to actuality are ipso facto incredibly, improbably fortunate.

These estimations are generally quite easy and often suggestive. For example, what is the volume of all the human blood in the world? The average adult male has about six quarts of blood, adult women slightly less, children considerably less. Thus, if we estimate that on average each of the approximately 5 billion people in the world has about one gallon of blood, we get about 5 billion (5×10^9) gallons of blood in the world. Since there are about 7.5 gallons per cubic foot, there are approximately 6.7×10^8 cubic feet of blood. The cube root of 6.7×10^8 is 870. Thus, all the blood in the world would fit into a cube 870 feet on a side, less than 1/200th of a cubic mile!

Central Park in New York has an area of 840 acres, or about 1.3 square miles. If walls were built about it, all the blood in the world would cover the park to a depth of something under 20 feet. The Dead Sea on the Israel–Jordan border has an area of 390 square miles. If all the world's blood were put into the Dead Sea, it would add only three-fourths of an inch to its depth. Even without any particular context, these figures are surprising; there isn't that much blood in the world! Compare this with the volume of all the grass, or of all the leaves, or of all the algae in the world, and man's marginal status among life forms, at least volume-wise, is vividly apparent.

Switching dimensions for a moment, consider the ratio of the speed of the supersonic Concorde, which travels about 2,000 miles per hour, to that of a snail, which moves 25 feet per hour, a pace equivalent to about .005 miles per hour. The Concorde's velocity is 400,000 times that of the snail. An even more impressive ratio is that between the speed with which an average computer adds ten-digit numbers and the rate at which human calculators do so. Computers perform this task more than a million times faster than we do with our snail-like scratchings, and for supercomputers the ratio is over a billion to one.

One last earthly calculation that a scientific consultant from M.I.T. uses to weed out prospective employees during job interviews: How long, he asks, would it take dump trucks to cart away an isolated mountain, say Japan's Mount Fuji, to ground level? Assume trucks come every fifteen minutes, twenty-four hours a day, are instantaneously filled with mountain dirt and rock, and leave without getting in each other's way. The answer's a little surprising and will be given later.

Gargantuan Numbers and the Forbes 400

A concern with scale has been a mainstay of world literature from the Bible to Swift's Lilliputians, from Paul Bunyan to Rabelais' Gargantua. Yet it's always struck me how inconsistent these various authors have been in their use of large numbers.

The infant Gargantua (whence "gargantuan") is said to have needed 17,913 cows to supply him with milk. As a young student he traveled to Paris on a mare that was as large as six elephants, and hung the bells of Notre Dame on the mare's neck as jingles. Returning home, he was attacked by cannon fire from a castle, and combed the cannonballs from his hair with a 900-foot-long rake. For a salad he cut lettuces as large as walnut trees, and devoured half a dozen pilgrims who'd hidden among the trees. Can you determine the internal inconsistencies of this story?

The book of Genesis says of the Flood that ". . . all the high hills that were under the whole heaven were covered . . ." Taken literally, this seems to indicate that there were 10,000 to 20,000 feet of water on the surface of the earth, equivalent to more than half a billion cubic miles of liquid! Since, according to biblical accounts, it rained for forty days and forty nights, or for only 960 hours, the rain must have fallen at a rate of at least fifteen feet per hour, certainly enough to sink any aircraft carrier, much less an ark with thousands of animals on board.

Determining internal inconsistencies such as these is one of the minor pleasures of numeracy. The point, however, is not that one should be perpetually analyzing numbers for their consistency and plausibility, but that, when necessary, information can be gleaned from the barest numerical facts, and claims can often be refuted on the basis of these raw numbers alone. If people were more capable of estimation and simple calculation, many obvious inferences would be drawn (or not), and fewer ridiculous notions would be entertained.

Before returning to Rabelais, let's consider two hanging wires of equal cross section. (This latter sentence, I'm sure, has never before appeared in print.) The forces on the wires are proportional to their masses, which are proportional to their lengths. Since the areas of the cross sections of the supporting wires are equal, the stresses in the wire, force divided by cross-sectional area, vary as the lengths of the wires. A wire ten times as long as another will have ten times the stress of the shorter one. Similar arguments show that, of two geometrically similar bridges of the same material, the larger one is necessarily the weaker of the two.

Likewise, a six-foot man cannot be scaled up to thirty feet, Rabelais notwithstanding. Multiplying his height by 5 will increase his weight by a factor of 5^3, while his ability to support this weight—as measured by the cross-sectional area of his bones—will increase by a factor of only 5^2. Elephants are big but at the cost of quite thick legs, while whales are relatively immune because they're submerged in water.

Although a reasonable first step in many cases, scaling quantities up or down proportionally is often invalid, as more mundane examples also demonstrate. If the price of bread goes up 6 percent, that's no reason to suspect the price of yachts will go up by 6 percent as well. If a company grows to twenty times its original size, the relative proportions of its departments will not stay the same. If ingesting a thousand grams of some substance causes one rat in one hundred to develop cancer, that's no guarantee that ingesting just one hundred grams will cause one rat in one thousand to develop cancer.

I once wrote to a significant minority of the Forbes 400, a list of the four hundred richest Americans, asking for $25,000 in support for a project I was working on at the time. Since the average wealth of the people I contacted was approximately $400 million ($4 \times 10^8$, certainly a gargantuan number of dollars) and I was asking for only 1/16,000th of that wealth, I hoped that linear proportionality would hold, reasoning that if some stranger wrote me asking for support of a worthy project of his and asked me for $25, more than 1/16,000th of my own net worth, I would probably comply with his request. Alas, though I received a number of kind responses, I didn't receive any money.

Archimedes and Practically Infinite Numbers

There is a fundamental property of numbers named after the Greek mathematician Archimedes which states that any number, no matter how huge, can be exceeded by adding together sufficiently many of any smaller number, no matter how tiny. Though obvious in principle, the consequences are sometimes resisted, as they were by the student of mine who maintained that human hair just didn't grow in miles per hour. Unfortunately, the nanoseconds used up in a simple computer operation do add up to lengthy bottlenecks on intractable problems, many of which would require millennia to solve in general. It takes some getting accustomed to the fact that the minuscule times and distances of microphysics as well as the vastness of astronomical phenomena share the dimensions of our human world.

It's clear how the above property of numbers led to Archimedes' famous pronouncement that given a fulcrum, a long enough lever, and a place to stand, he alone could physically lift the earth. An awareness of the additivity of small quantities is lacking in innumerates, who don't seem to believe that their little aerosol cans of hairspray could play any role in the depletion of the ozone layer of the atmosphere, or that their individual automobile contributes anything to the problem of acid rain.

The pyramids, impressive as they are, were built a stone at a time over a period very much shorter than the five thousand to ten thousand years required to move the 12,000-foot Mount Fuji by truck. A similar but more classic calculation of this type was made by Archimedes, who estimated the number of grains of sand needed to fill up the earth and heavens. Though he didn't have exponential notation, he invented something comparable, and his calculations were essentially equivalent to the following.

Interpreting "the earth and heavens" to be a sphere about the earth, we observe that the number of grains of sand needed to fill it depends on the radius of the sphere and the coarseness of the sand. Assuming there are fifteen grains per linear inch, there are 15×15 per planar inch and 15^3 grains per cubic inch. Since there are twelve inches per foot, there are 12^3 cubic inches per cubic foot and thus $15^3 \times 12^3$ grains per cubic foot. Similarly, there are $15^3 \times 12^3 \times 5,280^3$ grains per cubic mile. Since the formula for the volume of a sphere is $\frac{4}{3} \times$ pi $\times$ the cube of the radius, the number of grains of sand needed to fill a sphere of radius one trillion miles (approximately Archimedes' estimate) is $\frac{4}{3} \times$ pi $\times 1,000,000,000,000^3 \times 15^3 \times 12^3 \times 5,280^3$. This equals approximately 10^{54} grains of sand.

There is a sense of power connected with such calculations which is hard to explain but which somehow involves a mental encompassing of the world. A more modern version is the calculation of the approximate number of subatomic bits that would fill up the universe. This number plays the role of "practical infinity" for computer problems which are solvable but only theoretically.

The size of the universe is, to be a little generous, a sphere about 40 billion light-years in diameter. To be even more generous and also to simplify the rough calculation, assume it's a cube 40 billion light-years on a side. Protons and neutrons are about 10^{-12} centimeters in diameter. The Archimedean question computer scientist Donald Knuth poses is how many little cubes 10^{-13} centimeters in diameter (¹⁄₁₀ the diameter of these nucleons) would fit into the universe. An easy calculation shows the number to be less than 10^{125}. Thus, even if a computer the size of the universe had working parts that were smaller than nucleons, it would contain fewer than 10^{125} such parts, and thus computations on problems which require more parts wouldn't be possible. Perhaps surprisingly, there are many such problems, some of them quite ordinary and of practical importance.

A comparably tiny time unit is the amount of time required for light, which travels at 300,000 kilometers per second, to traverse the length of one of the above tiny cubes, whose edges are 10^{-13} centimeters. Taking the universe to be about 15 billion years old, we determine that fewer than 10^{42} such time units have passed since the beginning of time. Thus, any computer calculation which requires more than 10^{42} steps (each of which is certainly going to require more time than our unit of time) requires more time than the present history of this universe. Again, there are many such problems.

Taking a human being to be spherical and about a meter in diameter (assume a person is squatting), we end with some biologically revealing comparisons that are somewhat easier to visualize. The size of a human cell is to that of a person as a person's size is to that of Rhode Island. Likewise, a virus is to a person as a person is to the earth; an atom is to a person as a person is to the earth's orbit around the sun; and a proton is to a person as a person is to the distance to Alpha Centauri.

The Multiplication Principle and Mozart's Waltzes

Now is probably a good time to reiterate my earlier remark that an occasional difficult passage may be safely ignored by the innumerate reader. The next few sections in particular may contain several such passages. The occasional trivial passage likewise may be quite safely ignored by the numerate reader. (Indeed, the whole book may be safely ignored by all readers, but I'd prefer that, at most, only isolated paragraphs will be.)

The so-called multiplication principle is deceptively simple and very important. It states that if some choice can be made in M different ways and some subsequent choice can be made in N different ways, then there are M × N different ways these choices can be made in succession. Thus, if a woman has five blouses and three skirts, then she has 5 × 3 = 15 choices of outfit, since each of the five blouses (B1, B2, B3, B4, B5) can be worn with any of the three skirts (S1, S2, S3), to yield the following fifteen outfits: B1,S1; B1,S2; B1,S3; B2,S1; B2,S2; B2,S3; B3,S1; B3,S2; B3,S3; B4,S1; B4,S2; B4,S3; B5,S1; B5,S2; B5,S3. From a menu with four appetizers, seven entrees, and three desserts, a diner can design 4 × 7 × 3 = 84 different dinners, assuming he orders each course.

Likewise, the number of possible outcomes when rolling a pair of dice is 6 × 6 = 36; any of the six numbers on the first die can be combined with any of the six numbers on the second die. The number of possible outcomes where the second die differs from the first is 6 × 5 = 30; any of the six numbers of the first die can be combined with any of the remaining five numbers on the second die. The number of possible outcomes when rolling three dice is 6 × 6 × 6 = 216. The number of outcomes in which the numbers on the three dice differ is 6 × 5 × 4 = 120.

The principle is invaluable in calculating large numbers, such as the total number of telephones reachable without dialing an area code, which comes to roughly 8×10^6, or 8 million. The first position can be filled by any one of eight digits (0 and 1 aren't generally used in the first position), the second position by any one of the ten digits, and so on, up to the seventh position. (There are actually a few more constraints on the numbers and the positions they can fill, which brings the 8 million figure down somewhat.) Similarly, the number of possible license plates in a state whose plates all have two letters followed by four numbers is $26^2 \times 10^4$. If repetitions are not allowed, the number of possible plates is $26 \times 25 \times 10 \times 9 \times 8 \times 7$.

When the leaders of eight Western countries get together for the important business of a summit meeting—having their joint picture taken—there are $8 \times 7 \times 6 \times 5 \times 4 \times 3 \times 2 \times 1 = 40,320$ different ways in which they can be lined up. Why? Out of these 40,320 ways, in how many would President Reagan and Prime Minister Thatcher be standing next to each other? To answer this, assume that Reagan and Thatcher are placed in a large burlap bag. These seven entities (the six remaining leaders and the bag) can be lined up in $7 \times 6 \times 5 \times 4 \times 3 \times 2 \times 1 = 5,040$ ways (invoking the multiplication principle once again). This number must then be multiplied by two since, once Reagan and Thatcher are removed from the bag, we have a choice as to which one of the two adjacently placed leaders should be placed first. There are thus 10,080 ways for the leaders to line up in which Reagan and Thatcher are standing next to each other. Hence, if the leaders were randomly lined up, the probability that these two would be standing next to each other is $10,080/40,320 = \frac{1}{4}$.

Mozart once wrote a waltz in which he specified eleven different possibilities for fourteen of the sixteen bars of the waltz and two possibilities for one of the other bars. Thus, there are 2×11^{14} variations on the waltz, only a minuscule fraction of which have ever been heard. In a similar vein, the French poet Raymond Queneau once published a book entitled *Cent mille milliards de poèmes*, which consisted of a sonnet on each of ten pages. The pages were cut to allow each of the fourteen lines of each sonnet to be turned separately, so that any of the ten first lines could be combined with any of the ten second lines, and so on. Queneau claimed that all the resulting 10^{14} sonnets made sense, although it's safe to say that the claim will never be verified.

People don't generally appreciate how large such seemingly

tidy collections can be. A sportswriter once recommended in print that a baseball manager should play every possible combination of his twenty-five-member team for one game to find the nine that play best together. There are various ways to interpret this suggestion, but in all of them the number of games is so large that the players would be long dead before the games were completed.

Triple-Scoop Cones and Von Neumann's Trick

Baskin-Robbins ice-cream parlors advertise thirty-one different flavors of ice cream. The number of possible triple-scoop cones without any repetition of flavors is therefore 31 × 30 × 29 = 26,970; any of the thirty-one flavors can be on top, any of the remaining thirty in the middle, and any of the remaining twenty-nine on the bottom. If we're not interested in how the flavors are arranged on the cone but merely in how many three-flavored cones there are, we divide 26,970 by 6, to get 4,495 cones. The reason we divide by 6 is that there are 6 = 3 × 2 × 1 different ways to arrange the three flavors in, say, a strawberry-vanilla-chocolate cone: SVC, SCV, VSC, VCS, CVS, and CSV. Since the same can be said for any three-flavored cone, the number of such cones is (31 × 30 × 29)/(3 × 2 × 1) = 4,495.

A less fattening example is provided by the many state lotteries which require the winner to choose six numbers out of a possible forty. If we're concerned with the order in which these six numbers are chosen, then there are (40 × 39 × 38 × 37 × 36 × 35) = 2,763,633,600 ways of choosing them. If, however, we are interested only in the six numbers as a collection (as we are in the case of the lotteries) and not in the order in which they are chosen, then we divide 2,763,633,600 by 720 to determine the number of such collections: 3,838,380. The division is necessary since there are 720 = 6 × 5 × 4 × 3 × 2 × 1 ways to arrange the six numbers in any collection.

Another example, and one of considerable importance to

card players, is the number of possible five-card poker hands. There are 52 × 51 × 50 × 49 × 48 possible ways to be dealt five cards if the order of the cards dealt is relevant. Since it's not, we divide the product by (5 × 4 × 3 × 2 × 1), and find that there are 2,598,960 possible hands. Once that number is known, several useful probabilities can be computed. The chances of being dealt four aces, for example, is 48/2,598,960 (= about 1 in 50,000), since there are forty-eight possible ways of being dealt a hand with four aces corresponding to the forty-eight cards which could be the fifth card in such a hand.

Note that the form of the number obtained is the same in all three examples: (32 × 30 × 29)/(3 × 2 × 1) different three-flavored ice-cream cones; (40 × 39 × 38 × 37 × 36 × 35)/ (6 × 5 × 4 × 3 × 2 × 1) different ways to choose six numbers out of forty; and (52 × 51 × 50 × 49 × 48)/(5 × 4 × 3 × 2 × 1) different poker hands. Numbers obtained in this way are called combinatorial coefficients. They arise when we're interested in the number of ways of choosing R elements out of N elements and we're not interested in the order of the R elements chosen.

An analogue of the multiplication principle can be used to calculate probabilities. If two events are independent in the sense that the outcome of one event has no influence on the outcome of the other, then the probability that they both occur is computed by multiplying the probabilities of the individual events.

For example, the probability of obtaining two heads in two flips of a coin is ½ × ½ = ¼ since of the four equally likely possibilities—tail,tail; tail,head; head,tail; head,head—one is a pair of heads. For the same reason, the probability of five straight coin flips resulting in heads is $(½)^5 = $ ⅟₃₂ since one of the thirty-two equally likely possibilities is five consecutive heads.

Since the probability that a roulette wheel will stop on red is ¹⁸⁄₃₈, and since spins of a roulette wheel are independent, the probability the wheel will stop on red for five consecutive spins is $(^{18}/_{38})^5$ (or .024 – 2.4%). Similarly, given that the probability that someone chosen at random was not born in July is $^{11}/_{12}$, and given that people's birthdays are independent, the probability that none of twelve randomly selected people was born in July is $(^{11}/_{12})^{12}$ (or .352 – 35.2%). Independence of events is a very important notion in probability, and when it holds, the multiplication principle considerably simplifies our calculations.

One of the earliest problems in probability was suggested to the French mathematician and philosopher Pascal by the gambler Antoine Gombaud, Chevalier de Mère. De Mère wished to know which event was more likely: obtaining at least one 6 in four rolls of a single die, or obtaining at least one 12 in twenty-four rolls of a pair of dice. The multiplication principle for probabilities is sufficient to determine the answer if we remember that the probability that an event doesn't occur is equal to 1 minus the probability that it does (a 20 percent chance of rain implies an 80 percent chance of no rain).

Since $\frac{5}{6}$ is the probability of not rolling a 6 on a single roll of a die, $(\frac{5}{6})^4$ is the probability of not rolling a 6 in four rolls of the die. Hence, subtracting this number from 1 gives us the probability that this latter event (no 6s) doesn't occur; in other words, of there being at least one 6 rolled in the four tries: $1 - (\frac{5}{6})^4 = .52$. Likewise, the probability of rolling at least one 12 in twenty-four rolls of a pair of dice is seen to be $1 - (^{35}/_{36})^{24} = .49$.

A more contemporary instance of the same sort of calculation involves the likelihood of acquiring AIDS heterosexually. It's estimated that the chance of contracting AIDS in a single unprotected heterosexual episode from a partner known to have the disease is about one in five hundred (the average of the figures from a number of studies). Thus, the probability of not getting it from a single such encounter is 499/500. If these risks are independent, as many assume them to be, then the chances of not falling victim after two such encounters is $(499/500)^2$, and after N such encounters $(499/500)^N$. Since $(499/500)^{346}$ is $\frac{1}{2}$, one runs about a 50 percent chance of not contracting AIDS by having unsafe heterosexual intercourse every day for a year with someone who has the disease (and thus, equivalently, a 50 percent chance of contracting it).

With a condom the risk of being infected from a single unsafe heterosexual episode with someone known to have the disease falls to one in five thousand, and safe sex every day for ten years with such a person (assuming the victim's survival) would lead to a 50 percent chance of getting the disease yourself. If your partner's disease status is not known, but he or she is not a member of any known risk group, the chance per episode of contracting the infection is one in five million unprotected, one in fifty million with a condom. You're more likely to die in a car crash on the way home from such a tryst.

Two opposing parties often decide an outcome by the flip of a coin. One or both of the parties may suspect the coin is biased. A cute little trick utilizing the multiplication principle was devised by mathematician John von Neumann to allow the contestants to use the biased coin and still get fair results.

The coin is flipped twice. If it comes up heads both times or tails both times, it is flipped twice again. If it comes up heads-tails, this will decide the outcome in favor of the first party, and if it comes up tails-heads, this will decide the outcome in favor of the second party. The probabilities of both these outcomes are the same even if the coin is biased. For example, if the coin lands heads 60 percent of the time and tails 40 percent of the time, a heads-tails sequence has probability $.6 \times .4 = .24$ and a tails-heads sequence has probability $.4 \times .6 = .24$. Thus, both parties can be confident of the fairness of the outcome despite the possible bias of the coin (unless it is crooked in some different way).

An important bit of background intimately connected to the multiplication principle and combinatorial coefficients is the binomial probability distribution. It arises whenever a procedure or trial may result in "success" or "failure" and one is interested in the probability of obtaining R successes in N trials. If 20 percent of all sodas dispensed by a vending machine overflow their cups, what is the probability that exactly three of the next ten will overflow? at most, three? If a family has five children, what is the probability that they will have exactly three girls? at least, three? If one-tenth of all people have a certain blood type, what is the probability that, of the next hundred people we randomly select, exactly eight will have the blood type in question? at most, eight?

Let me derive the answer to the questions about the vending machine, 20 percent of whose sodas overflow their cups. The probability that the first three sodas overflow and the next seven do not is, by the multiplication principle for probability, $(.2)^3 \times (.8)^7$. But there are many different ways for exactly three of the ten cups to overflow, each way having probability $(.2)^3 \times (.8)^7$. It may be that only the last three cups overflow, or only the fourth, fifth, and ninth cups, and so on. Thus, since there are altogether $(10 \times 9 \times 8)/(3 \times 2 \times 1) = 120$ ways for us to pick three out of the ten cups (combinatorial coefficient), the probability of some collection of exactly three cups overflowing is $120 \times (.2)^3 \times (.8)^7$.

The probability of at most three cups overflowing is determined by finding the probability of exactly three cups overflowing, which we've done, and adding to it the probabilities of exactly two, one, and zero cups overflowing, which can be determined in a similar way. Happily, there are tables and good approximations which can be used to shorten these calculations.

Julius Caesar and You

Two final applications of the multiplication principle—one slightly depressing, the other somewhat cheering. The first is the probability of not being afflicted with any of a variety of diseases, accidents, or other misfortunes. Not being killed in a car accident may be 99 percent certain, while 98 percent of us may avoid perishing in a household accident. Our chances of escaping lung disease may be 95 percent; dementia, 90 percent; cancer, 80 percent; and heart disease, 75 percent. These figures are merely for illustration, but accurate estimates may be made for a wide range of dire possibilities. While the chances of avoiding any particular disease or accident may be encouraging, the probability of avoiding them all is not. If we multiply all the above probabilities (assuming these calamities are largely independent), the product grows disturbingly small quite quickly: already our chance of not suffering any of the few misfortunes listed above is less than 50 percent. It's a little anxiety-provoking, how this innocuous multiplication principle can make our mortality more vivid.

Now for better news of a kind of immortal persistence. First, take a deep breath. Assume Shakespeare's account is accurate and Julius Caesar gasped "You too, Brutus" before breathing his last. What are the chances you just inhaled a molecule which Caesar exhaled in his dying breath? The surprising answer is that, with probability better than 99 percent, you did just inhale such a molecule.

For those who don't believe me: I'm assuming that after more than two thousand years the exhaled molecules are uniformly spread about the world and the vast majority are still free in the atmosphere. Given these reasonably valid assumptions, the problem of determining the relevant probability is straightforward. If there are N molecules of air in the world and Caesar exhaled A of them, then the probability that any given molecule you inhale is from Caesar is A/N. The probability that any given molecule you inhale is not from Caesar is thus $1 - A/N$. By the multiplication principle, if you inhale three molecules, the probability that none of these three is from Caesar is $[1 - A/N]^3$. Similarly, if you inhale B molecules, the probability that none of them is from Caesar is approximately $[1 - A/N]^B$. Hence, the probability of the complementary event, of your inhaling at least one of his exhaled molecules, is $1 - [1 - A/N]^B$. A, B (each about 1/30th of a mole, or 2.2×10^{22}), and N (about 10^{44} molecules) are such that this probability is more than .99. It's intriguing that we're all, at least in this minimal sense, eventually part of one another.

2

Probability and Coincidence

It is no great wonder if, in the long process of time, while fortune takes her course hither and thither, numerous coincidences should spontaneously occur.   —Plutarch

"You're a Capricorn, too. That's so exciting."

A man who travels a lot was concerned about the possibility of a bomb on board his plane. He determined the probability of this, found it to be low but not low enough for him, so now he always travels with a bomb in his suitcase. He reasons that the probability of two bombs being on board would be infinitesimal.

Some Birthday vs. a Particular Birthday

Sigmund Freud once remarked that there was no such thing as a coincidence. Carl Jung talked about the mysteries of synchronicity. People in general prattle ceaselessly about ironies here and ironies there. Whether we call them coincidences, synchronicities, or ironies, however, these occurrences are much more common than most people realize.

Some representative examples: "Oh, my brother-in-law went to school there, too, and my friend's son cuts the principal's lawn, and my neighbor's daughter knows a girl who once was a cheerleader for the school." — "There've been five instances of the fish idea since this morning when she told me of her fears about his fishing on the open lake. Fish for lunch, the fish motif on Caroline's dress, the . . ." — Christopher Columbus discovered the New World in 1492 and his fellow Italian Enrico Fermi discovered the new world of the atom in 1942. — "You said you wanted to keep up with him, but later you said you wanted to keep abreast of her. It's clear what's on your mind." —The ratio of the height of the Sears Building in Chicago to the height of the Woolworth Building in New York is the same to four significant digits (1.816 vs. 1816) as the ratio of the mass of a proton to the mass of an electron. —The Reagan–Gorbachev INF treaty was signed on December 8, 1987, exactly seven years after John Lennon was killed.

A tendency to drastically underestimate the frequency of coincidences is a prime characteristic of innumerates, who generally accord great significance to correspondences of all sorts while attributing too little significance to quite conclusive but less flashy statistical evidence. If they anticipate someone else's thought, or have a dream that seems to come true, or read that, say, President Kennedy's secretary was named Lincoln while President Lincoln's secretary was named Kennedy, this is considered proof of some wondrous but mysterious harmony that somehow holds in their personal universe. Few experiences are more dispiriting to me than meeting someone who seems intelligent and open to the world but who immediately inquires about my zodiac sign and then begins to note characteristics of my personality consistent with that sign (whatever sign I give them).

The surprising likelihood of coincidence is illustrated by the following well-known result in probability. Since a year has 366 days (if you count February 29), there would have to be 367 people gathered together in order for us to be absolutely certain that at least two people in the group have the same birthday. Why?

Now, what if we were content to be just 50 percent certain

of this? How many people would there have to be in a group in
order for the probability to be half that at least two people in it
have the same birthday? An initital guess might be 183, about
half of 365. The surprising answer is that there need be only
twenty-three. Stated differently, fully half of the time that
twenty-three randomly selected people are gathered together,
two or more of them will share a birthday.

For readers unwilling to accept this on faith, here is a brief
derivation. By the multiplication principle, the number of ways
in which five dates can be chosen (allowing for repetitions) is
(365 × 365 × 365 × 365 × 365). Of all these 365^5 ways,
however, only (365 × 364 × 363 × 362 × 361) are such that
no two of the dates are the same; any of the 365 days can be
chosen first, any of the remaining 364 can be chosen second,
and so on. Thus, by dividing this latter product (365 × 364 ×
363 × 362 × 361) by 365^5, we get the probability that five
people chosen at random will have no birthday in common. Now,
if we subtract this probability from 1 (or from 100 percent if
we're dealing in percentages), we get the complementary probability that at least two of the five people do have a birthday in
common. A similar calculation using 23 rather than 5 yields ½,
or 50 percent, as the probability that at least two of twenty-three
people will have a common birthday.

A couple of years ago, someone on the Johnny Carson show
was trying to explain this. Johnny Carson didn't believe it, noted
that there were about 120 people in the studio audience, and
asked how many of them shared his birthday of, say, March 19.
No one did, and the guest, who wasn't a mathematician, said
something incomprehensible in his defense. What he should
have said is that it takes twenty-three people to be 50 percent
certain that there is *some* birthday in common, not any *particular*
birthday such as March 19. It requires a large number of people,
253 to be exact, to be 50 percent certain that someone in the
group has March 19 as his or her birthday.

A brief derivation of the last fact: Since the probability of
someone's birthday not being March 19 is 364/365, and since

birthdays are independent, the probability of two people not having March 19 as a birthday is 364/365 × 364/365. Thus, the probability of N people not having March 19 as a birthday is $(364/365)^N$, which, when N = 253, is approximately ½. Hence, the complementary probability that at least one of these 253 people was born on March 19 is also ½, or 50 percent.

The moral, again, is that some unlikely event is likely to occur, whereas it's much less likely that a particular one will. Martin Gardner, the mathematics writer, illustrates the distinction between general and specific occurrences by means of a spinner with the twenty-six letters of the alphabet on it. If the spinner is spun one hundred times and the letters recorded, the probability that the word CAT or WARM will appear is very small, but the probability of *some* word's appearing is high. Since I brought up the topic of astrology, Gardner's examples of the first letters of the names of the months and the planets are particularly appropriate. The months—JFMAMJJASOND—give us JASON; the planets—MVEMJSUNP—spell SUN. Significant? No.

The paradoxical conclusion is that it would be very unlikely for unlikely events not to occur. If you don't specify a predicted event precisely, there are an indeterminate number of ways for an event of that general kind to take place.

Medical quackery and television evangelism will be discussed in the next chapter, but it should be mentioned here that their predictions are usually sufficiently vague so that the probability of some event of the predicted kind occurring is very high; it's the particular predictions that seldom come true. That some nationally famous politician will undergo a sex-change operation, as a newspaper astrologer-psychic recently predicted, is considerably more likely than that New York's Mayor Koch will. That some viewer will be relieved of his gastric pains just as a television evangelist calls out the symptoms is considerably more likely than that a particular viewer will be. Likewise, insurance policies with broad coverage which compensates for any mishap are apt to be cheaper in the long run than insurance for a particular disease or a particular trip.

Chance Encounters

Two strangers from opposite sides of the United States sit next to each other on a business trip to Milwaukee and discover that the wife of one of them was in the tennis camp run by an acquaintance of the other's. This sort of coincidence is surprisingly common. If we assume each of the approximately 200 million adults in the United States knows about 1,500 people, and that these 1,500 people are reasonably spread out around the country, then the probability is about one in a hundred that they will have an acquaintance in common, and more than ninety-nine in a hundred that they will be linked by a chain of two intermediates.

We can be almost certain, then, given these assumptions, that two people chosen at random will be linked, as were the strangers on the business trip, by a chain of at most two intermediates. Whether they'll run down the 1,500 or so people they each know (as well as the acquaintances of each of these 1,500) during their conversation and thus become aware of the two intermediates linking them is another, more dubious matter.

These assumptions can be relaxed somewhat. Maybe the average adult knows fewer than 1,500 other adults, or, more likely, most of the people he or she does know live close by and are not spread about the country. Even in these cases, however, the probability of two randomly selected people being linked by two intermediates is unexpectedly high.

A more empirical approach to coincidental meetings was taken by psychologist Stanley Milgrim, who gave each member of a randomly selected group of people a document and a (different) "target individual" to whom the document was to be transmitted. The directions were that each person was to send the document to the person he knew who was most likely to know the target individual, and that he was to direct that person to do the same, until the target individual was reached. Milgrim found that the number of intermediate links ranged from two to ten, with five being the most common number. This study is more impressive, even if less spectacular, than the earlier a priori probability argument. It goes some way toward explaining how confidential information, rumors, and jokes percolate so rapidly through a population.

If the target is well known, the number of intermediates is even smaller, especially if you have a link with one or two celebrities. How many intermediates are there between you and President Reagan? Say the number is N. Then the number of intermediates between you and Secretary General Gorbachev is less than or equal to (N + 1), since Reagan has met Gorbachev. How many intermediates between you and Elvis Presley? Again, it can't be bigger than (N + 2), since Reagan's met Nixon, who's met Presley. Most people are surprised when they realize how short the chain is which links them to almost any celebrity.

When I was a freshman in college, I wrote a letter to English philosopher and mathematician Bertrand Russell telling him that he'd been an idol of mine since junior high school and asking him about something he'd written concerning the German philosopher Hegel's theory of logic. Not only did he answer my letter, but he included his response in his autobiography, sandwiched between letters to Nehru, Khrushchev, T. S. Eliot, D. H. Lawrence, Ludwig Wittgenstein, and other luminaries. I like to maintain that the number of intermediates linking me to these historical figures is one: Russell.

Another problem in probability illustrates how common coincidences may be in another context. The problem's often phrased in terms of a large number of men who check their hats at a restaurant, whereupon the attendant promptly scrambles the hat-check numbers randomly. What is the probability that at least one of the men will get his own hat upon leaving? It's natural to think that if the number of men is very large, this probability should be quite small. Surprisingly, about 63 percent of the time, at least one man will get his own hat back.

Put another way: If a thousand addressed envelopes and a thousand addressed letters are thoroughly scrambled and one letter is then placed into each envelope, the probability is likewise about 63 percent that at least one letter will find its way into its corresponding envelope. Or take two thoroughly shuffled decks of cards. If cards from each of these decks are turned over one at a time in tandem, what is the probability that at least one exact match will occur? Again, about 63 percent. (Peripheral question: Why is it necessary to shuffle only one of the decks thoroughly?)

A very simple numerical principle that's sometimes of use in accounting for the certainty of a particular kind of coincidence is illustrated by the mailman who has twenty-one letters to distribute among twenty mailboxes. Since 21 is greater than 20, he can be sure, even without looking at the addresses, that at least one mailbox will get more than one letter. This bit of common sense, sometimes termed the pigeonhole or Dirichlet drawer principle, can occasionally be used to derive claims that are not so obvious.

We invoked it in stating that if we had 367 people gathered together, we could be certain that at least two had the same birthday. A more interesting fact is that at least two people living in Philadelphia must have the same number of hairs on their heads. Consider the numbers up to 500,000, a figure that's generally taken to be an upper bound for the number of hairs on any human head, and imagine these numbers to be the labels on half a million mailboxes. Imagine further that each of the 2.2 million Philadelphians is a letter to be delivered to the mailbox whose label corresponds to the number of hairs on his or her head. Thus, if Mayor Wilson Goode has 223,569 hairs on his head, then he is to be delivered to the mailbox with that number.

Since 2,200,000 is considerably more than 500,000, we can be certain that at least two people have the same number of hairs on their heads; i.e., that some mailbox will receive at least two Philadelphians. (Actually, we can be sure that at least five Philadelphians have the same number of hairs on their heads. Why?)

A Stock-Market Scam

Stock-market advisers are everywhere, and you can probably find one to say almost anything you might want to hear. They're usually assertive, sound quite authoritative, and speak a strange language of puts, calls, Ginnie Maes, and zero-coupons. In my humble experience, most don't really know what they're talking about, but presumably some do.

If from some stock-market adviser you received in the mail for six weeks in a row correct predictions on a certain stock index and were asked to pay for the seventh such prediction, would you? Assume you really are interested in making an investment of some sort, and assume further that the question is being posed to you before the stock crash of October 19, 1987. If you would be willing to pay for the seventh prediction (or even if you wouldn't), consider the following con game.

Some would-be adviser puts a logo on some fancy stationery and sends out 32,000 letters to potential investors in a stock index. The letters tell of his company's elaborate computer model, his financial expertise and inside contacts. In 16,000 of these letters he predicts the index will rise, and in the other 16,000 he predicts a decline. No matter whether the index rises or falls, a follow-up letter is sent, but only to the 16,000 people who initially received a correct "prediction." To 8,000 of them, a rise is predicted for the next week; to the other 8,000, a decline. Whatever happens now, 8,000 people will have received two correct predictions. Again, to these 8,000 people only, letters are sent concerning the index's performance the following week: 4,000 predicting a rise; 4,000, a decline. Whatever the outcome, 4,000 people have now received three straight correct predictions.

This is iterated a few more times, until 500 people have received six straight correct "predictions." These 500 people are now reminded of this and told that in order to continue to receive this valuable information for the seventh week they must each contribute $500. If they all pay, that's $250,000 for our adviser.

If this is done knowingly and with intent to defraud, this is an illegal con game. Yet it's considered acceptable if it's done unknowingly by earnest but ignorant publishers of stock newsletters, or by practitioners of quack medicine, or by television evangelists. There's always enough random success to justify almost anything to someone who wants to believe.

There is another quite different problem exemplified by these stock-market forecasts and fanciful explanations of success. Since they're quite varied in format and often incomparable and very numerous, people can't act on all of them. The people who try their luck and don't fare well will generally be quiet about their experiences. But there'll always be some people who will do extremely well, and they will loudly swear to the efficacy of whatever system they've used. Other people will soon follow suit, and a fad will be born and thrive for a while despite its baselessness.

There is a strong general tendency to filter out the bad and the failed and to focus on the good and the successful. Casinos encourage this tendency by making sure that every quarter that's won in a slot machine causes lights to blink and makes its own little tinkle in the metal tray. Seeing all the lights and hearing all the tinkles, it's not hard to get the impression that everyone's winning. Losses or failures are silent. The same applies to well-publicized stock-market killings vs. relatively invisible stock-market ruinations, and to the faith healer who takes credit for any accidental improvement but will deny responsibility if, for example, he ministers to a blind man who then becomes lame.

This filtering phenomenon is very widespread and manifests itself in many ways. Along almost any dimension one cares to choose, the average value of a large collection of measurements is about the same as the average value of a small collection, whereas the extreme value of a large collection is considerably more extreme than that of a small collection. For example, the average water level of a given river over a twenty-five-year period will be approximately the same as the average water level over a one-year period, but the worst flood over a twenty-five-year period is apt to be considerably higher than that over a one-year period. The average scientist in tiny Belgium will be comparable to the average scientist in the United States, even though the best scientist in the United States will in general be better than Belgium's best (we ignore obvious complicating factors and definitional problems).

So what? Because people usually focus upon winners and extremes whether they be in sports, the arts, or the sciences, there's always a tendency to denigrate today's sports figures, artists, and scientists by comparing them with extraordinary cases. A related consequence is that international news is usually worse than national news, which in turn is usually worse than state news, which is worse than local news, which is worse than the news in your particular neighborhood. Local survivors of tragedy are invariably quoted on TV as saying something like, "I can't understand it. Nothing like that has ever happened around here before."

One final manifestation: Before the advent of radio, TV, and film, musicians, athletes, etc., could develop loyal local audiences since they were the best that most of these people would ever see. Now audiences, even in rural areas, are no longer as satisfied with local entertainers and demand world-class talent. In this sense, these media have been good for audiences and bad for performers.

Expected Values: From Blood Testing to Chuck-a-Luck

Coincidences or extreme values catch the eye, but average or "expected" values are generally more informative. The expected value of a quantity is simply the average of its values weighted according to their probabilities. For example, if $\frac{1}{4}$ of the time a quantity equals 2, $\frac{1}{3}$ of the time it equals 6, another $\frac{1}{3}$ of the time it equals 15, and the remaining $\frac{1}{12}$ of the time it equals 54, then its expected value equals 12. This is so since $[12 = (2 \times \frac{1}{4}) + (6 \times \frac{1}{3}) + (15 \times \frac{1}{3}) + (54 \times \frac{1}{12})]$.

As a simple illustration, consider a home-insurance company. Assume it has good reason to believe that, on average, each year one out of every 10,000 of its policies will result in a claim of $200,000; one out of 1,000 policies will result in a claim of $50,000; one out of 50 will result in a claim of $2,000; and the remainder will result in a claim of $0. The insurance company would like to know what its average payout is per policy written. The answer is the expected value, which in this case is ($200,000 × 1/10,000) + ($50,000 × 1/1,000) + ($2,000 × 1/50) + ($0 × 9,789/10,000) = $20 + $50 + $40 = $110.

The expected payout on a slot machine is determined in like manner. Each payout is multiplied by the probability of its occurring, and these products are then summed up to give the average or expected payout. For example, if cherries on all three dials result in a payout of $80 and the probability of this is $(\frac{1}{20})^3$ (assume there are twenty entries on each dial, only one of which is a cherry), we multiply $80 by $(\frac{1}{20})^3$ and then add to this product the products of the other payouts (a loss being considered a negative payout) and their probabilities.

An illustration which isn't quite so vanilla: Assume a medical clinic tests blood for a certain disease from which approximately one person in a hundred suffers. People come to the clinic in groups of fifty, and the director wonders whether, instead of testing them individually, he should pool the fifty samples and test them all together. If the pooled sample is negative, he could pronounce the whole group healthy, and if not, he could then test each person individually. What is the expected number of tests the director will have to perform if he pools the blood samples?

The director will have to perform either one test (if the pooled sample is negative) or fifty-one tests (if it's positive). The probability that any one person is healthy is $\frac{99}{100}$, and so the probability that all fifty people are healthy is $(\frac{99}{100})^{50}$. Thus, the probability that he'll have to perform just one test is $(\frac{99}{100})^{50}$. On the other hand, the probability that at least one person suffers from the disease is the complementary probability $[1 - (\frac{99}{100})^{50}]$, and so the probability of having to perform fifty-one tests is

$[1 - (^{99}/_{100})^{50}]$. Thus, the expected number of tests necessary is $(1 \text{ test} \times (^{99}/_{100})^{50}) + (51 \text{ tests} \times [1 - (^{99}/_{100})^{50}]) = $ approximately 21 tests.

If there are large numbers of people having the blood test, the medical director would be wise to take part of each sample, pool it, and test this pooled sample first. If necessary, he could then test the remainders of each of the fifty samples individually. On average, this would require only twenty-one tests to test fifty people.

An understanding of expected values is helpful in analyzing most casino games, as well as the lesser-known game of chuck-a-luck which is played at carnivals in the Midwest and England.

The spiel that goes with chuck-a-luck can be very persuasive. You pick a number from 1 to 6 and the operator rolls three dice. If the number you pick comes up on all three dice, the operator pays you $3; if it comes up on two of the three dice, he pays you $2; and if it comes up on just one of the three dice, he pays you $1. Only if the number you picked doesn't come up at all do you pay him anything—just $1. With three different dice, you have three chances to win, and furthermore you'll sometimes win more than $1, while that is your maximum loss.

As Joan Rivers might say, "Can we calculate?" (If you'd rather not calculate, skip to the end of this section.) The probability of your winning is clearly the same no matter what number you choose, so, to make the calculation specific, assume you always pick the number 4. Since the dice are independent, your chances that a 4 will come up on all three dice are $^{1}/_{6} \times ^{1}/_{6} \times ^{1}/_{6} = ^{1}/_{216}$; so, approximately 1/216th of the time you'll win $3.

Your chances of a 4 coming up only twice are a little harder to calculate unless you use the binomial probability distribution mentioned in Chapter 1, which I'll derive again in this context. A 4 coming up on two of the three dice can happen in three different and mutually exclusive ways: X44, 4X4, or 44X, the X indicating a non-4. The probability of the first is $^{5}/_{6} \times ^{1}/_{6} \times ^{1}/_{6} = ^{5}/_{216}$, a result which holds true for the second and third ways as well. Adding, we find that the probability of a 4 coming up on two of the three dice is $15/216$, which is the fraction of the time you'll win $2.

The probability of obtaining exactly one 4 among the three dice is likewise determined by breaking the event into the three mutually exclusive ways it can happen. The probability of obtaining 4XX is $1/6 \times 5/6 \times 5/6 = 25/216$, which is also the probability of obtaining X4X and XX4. Adding, we get $75/216$ for the probability of exactly one 4 coming up on the three dice, and hence for the probability of your winning $1. To find the probability that no 4s come up when we roll three dice, we find how much probability is left over. That is, we subtract $(1/216 + 15/216 + 75/216)$ from 1 (or 100%) to get $125/216$. Thus, on the average, 125 out of 216 times you play chuck-a-luck, you'll lose $1.

The expected value of your winnings is thus $(\$3 \times 1/216) + (\$2 \times 15/216) + (\$1 \times 75/216) + (-\$1 \times 125/216) = \$(-17/216) = -\$.08$, and so, on the average, you would lose approximately eight cents every time you played this seemingly attractive game.

Choosing a Spouse

There are two approaches to love—through the heart and through the head. Neither one seems to work very well alone, but together . . . they still don't work too well. Nevertheless, there's probably a better chance of success if both are used. Upon thinking of past loves, someone who approaches romance through the heart is likely to bemoan lost opportunities and conclude that he or she will never again love as deeply. Someone who takes a more hardheaded approach may be interested in the following result in probability.

The model we'll consider assumes that our heroine—call her Myrtle—has reason to believe that she'll meet N potential spouses (spice?) during her "dating life." N could be two for some women, two hundred for others. The question Myrtle poses to herself is: When should I accept Mr. X and forgo the suitors who would come after him, some of whom may possibly be "better" than he? We'll assume she meets men sequentially, can judge the relative suitability for her of those she's met, and once she's rejected someone, he's gone forever.

For illustration, suppose Myrtle has met six men so far and that she rates them as follows: 3 5 1 6 2 4. That is, of the six men she's met, she liked the first one she met third-best, the second one she liked fifth-best, the third one she liked best of all, and so on. If the seventh man she meets she prefers to everyone except her favorite, her updated ranking would become: 4 6 1 7 3 5 2. After each man, she updates her relative ranking of her suitors and wonders what rule she should follow in order to maximize her chances of choosing the best of her projected N suitors.

The derivation of the best policy uses the idea of conditional probability (which we'll introduce in the next chapter) and a little calculus. The policy itself, though, is quite simple to describe. Call a suitor a heartthrob if he's better than all previous candidates. Myrtle should reject approximately the first 37 percent of the N candidates she's likely to meet, and then accept the first suitor after that (if any) who is a heartthrob.

For instance, suppose Myrtle isn't overly attractive and is likely to meet only four eligible suitors, and suppose further that these four men are equally likely to come to her in any of the twenty-four possible orderings ($24 = 4 \times 3 \times 2 \times 1$).

Since 37 percent is between 25 percent and 50 percent, the policy is ambiguous here, but the two best strategies correspond to the following: (A) Pass up the first candidate (25 percent of N = 4) and accept the first heartthrob after that. (B) Pass up the first two candidates (50 percent of N = 4) and accept the first heartthrob after that. Strategy A will result in Myrtle's choosing the best suitor in eleven of the twenty-four instances, while strategy B will result in success in ten of the twenty-four instances.

The list of all such sequential orderings is below, with the number 1 representing, as before, the suitor Myrtle would most prefer, 2 her second choice, etc. Thus, the ordering 3 2 1 4 indicates that she meets her third choice first, her second choice second, her first choice she meets third, and her last choice last. The orderings are marked with an A or a B to indicate in which instances these strategies result in her getting her first choice.

   1234 · 1243 · 1324 · 1342 · 1423 · 1432 · 2134(A) ·
   2143(A) · 2314(A,B) · 2341(A,B) · 2413(A,B) ·
   2431(A,B) · 3124(A) · 3142(A) · 3214(B) · 3241(B) ·
   3412(A,B) · 3421 · 4123(A) · 4132(A) · 4213(B) ·
   4231(B) · 4312(B) · 4321

   If Myrtle is quite attractive and can expect to have twenty-five suitors, her best strategy would still be to reject the first nine of these suitors (37 percent of 25) and then accept the first heartthrob after that. This could be verified by tabulation, as above, but the tables get unwieldy and it's best to accept the general proof. (Needless to say, the same analysis holds if the person seeking a spouse is a Mortimer and not a Myrtle.)

   For large values of N, the probability that Myrtle will find her Mr. Right following this 37 percent rule is also approximately 37 percent. Then comes the hard part: living with Mr. Right. Variants of the model exist with more romantically plausible constraints.

Coincidence and the Law

   In 1964 in Los Angeles a blond woman with a ponytail snatched a purse from another woman. The thief fled on foot but was later spotted entering a yellow car driven by a black man with a beard and a mustache. Police investigation eventually discovered a blond woman with a ponytail who regularly associated with a bearded and mustachioed black man who owned a yellow car. There wasn't any hard evidence linking the couple to the crime, or any witnesses able to identify either party. There was, however, agreement on the above facts.

The prosecutor argued that the probability was so low that such a couple existed that the police investigation must have turned up the actual culprits. He assigned the following probabilities to the characteristics in question: yellow car—1/10; man with a mustache—1/4; woman with a ponytail—1/10; woman with blond hair—1/3; black man with a beard—1/10; interracial couple in a car—1/1,000. The prosecutor further argued that the characteristics were independent, so that the probability that a randomly selected couple would have all of them would be $1/10 \times 1/4 \times 1/10 \times 1/3 \times 1/10 \times 1/1,000 = 1/12,000,000$, a number so low the couple must be guilty. The jury convicted them.

The case was appealed to the California supreme court, where it was overturned on the basis of another probability argument. The defense attorney in that trial argued that 1/12,000,000 was not the relevant probability. In a city the size of Los Angeles, with maybe 2,000,000 couples, the probability was not that small, he maintained, that there existed more than one couple with that particular list of characteristics, given that there was at least one such couple—the convicted couple. On the basis of the binomial probability distribution and the 1/12,000,000 figure, this probability can be determined to be about 8 percent—small, but certainly allowing for reasonable doubt. The California supreme court agreed and overturned the earlier guilty verdict.

Whatever the problems of the one in 12,000,000 figure, rarity by itself shouldn't necessarily be evidence of anything. When one is dealt a bridge hand of thirteen cards, the probability of being dealt that particular hand is less than one in 600 billion. Still, it would be absurd for someone to be dealt a hand, examine it carefully, calculate that the probability of getting it is less than one in 600 billion, and then conclude that he must not have been dealt that very hand because it is so very improbable.

In some contexts, improbabilities are to be expected. Every bridge hand is quite improbable. Likewise with poker hands or lottery tickets. In the case of the California couple, improbability carries more weight, but still, their defense attorney's argument was the right one.

Why is it, incidentally, if all the 3,838,380 ways of choosing six numbers out of forty are equally likely, that a lottery ticket with the numbers 2 13 17 20 29 36 is for most people much preferable to one with the numbers 1 2 3 4 5 6? This is, I think, a fairly deep question.

The following sports anomaly has legal implications as well. Consider two baseball players, say, Babe Ruth and Lou Gehrig. During the first half of the season, Babe Ruth hits for a higher batting average than Lou Gehrig. And during the second half of the season, Babe Ruth again hits for a higher average than Lou Gehrig. But for the season as whole, Lou Gehrig has a higher batting average than Babe Ruth. Could this be the case? Of course, the mere fact that I pose the question may cause some misgivings, but at first glance such a situation seems impossible.

What can happen is that Babe Ruth could hit .300 the first half of the season and Lou Gehrig only .290, but Ruth could bat two hundred times to Gehrig's one hundred times. During the second half of the season, Ruth could bat .400 and Gehrig only .390, but Ruth could come to bat only a hundred times to Gehrig's two hundred times at bat. The result would be a higher overall batting average for Gehrig than for Ruth: .357 vs. .333. You can't average batting averages.

There was an intriguing discrimination case in California several years ago which had the same formal structure as this batting-average puzzle. Looking at the proportion of women in graduate school at a large university, some women filed a lawsuit claiming that they were being discriminated against by the graduate school. When administrators sought to determine which departments were most guilty, they found that in each department a higher percentage of women applicants were admitted than men applicants. Women, however, applied in disproportionately large numbers to departments such as English and psychology that admitted only a small percentage of their applicants, whereas men applied in disproportionately large numbers to departments such as mathematics and engineering that admitted a much higher percentage of their applicants. The men's admissions pattern was analogous to Gehrig's hitting pattern—coming to bat more often during the second half of the season when getting a hit is easier.

Another counter-intuitive problem involving seemingly disproportionate probabilities concerns a New York City man who has a woman friend in the Bronx and one in Brooklyn. He is equally attached (or perhaps unattached) to each of them and thus is indifferent to whether he catches the northbound subway to the Bronx or the southbound subway to Brooklyn. Since both trains run every twenty minutes throughout the day, he figures he'll let the subway decide whom he'll visit, and take the first train which comes along. After a while, though, his Brooklyn woman friend, who's enamored of him, begins to complain that he shows up for only about one-fourth of his dates with her, while his Bronx friend, who's getting sick of him, begins to complain that he appears for three-fourths of his dates with her. Aside from callowness, what is this man's problem?

The simple answer follows, so don't read on if you want to think a bit. The man's more frequent trips to the Bronx are a result of the way the trains are scheduled. Even though they each come every twenty minutes, the schedule may be something like the following: Bronx train, 7:00; Brooklyn train, 7:05; Bronx train, 7:20; Brooklyn train, 7:25; and so on. The gap between the last Brooklyn train and the next Bronx train is fifteen minutes, three times as long as the five-minute gap between the last Bronx train and the next Brooklyn train, and thus accounts for his showing up for three-fourths of his dates in the Bronx and only for one-fourth of his Brooklyn dates.

Countless similar oddities result from our conventional ways of measuring, reporting, and comparing periodic quantities, whether they be the monthly cash flow of a government or the regular daily fluctuations in body temperature.

Fair Coins and Life's Winners and Losers

Imagine flipping a coin many times in succession and obtaining some sequence of heads and tails; say, HHTHTTTHH THTTTHTTHHHTHTTHHTHHTTHTHHTTHHTHT HHHHTHHHTT. If the coin is fair, there are a number of extremely odd facts about such sequences. For example, if one were to keep track of the proportion of the time that the number of heads exceeded the number of tails, one might be surprised that it is rarely close to half.

Imagine two players, Peter and Paul, who flip a coin at the rate of once a day and who bet on heads and tails respectively. Peter is ahead at any given time if there've been more heads up until then, while Paul is ahead if there've been more tails. Peter and Paul are each equally likely to be ahead at any given time, but whoever is ahead will probably have been ahead almost the whole time. If there have been one thousand coin flips, then if Peter is ahead at the end, the chances are considerably greater that he's been ahead more than 90 percent of the time, say, than that he's been ahead between 45 percent and 55 percent of the time! Likewise, if Paul is ahead at the end, it's considerably more likely that he's been ahead more than 96 percent of the time than that he's been ahead between 48 percent and 52 percent of the time.

Perhaps the reason this result is so counter-intuitive is that most people tend to think of deviations from the mean as being somehow bound by a rubber band: the greater the deviation, the greater the restoring force toward the mean. The so-called gambler's fallacy is the mistaken belief that because a coin has come up heads several times in a row, it's more likely to come up tails on its next flip (similar notions hold for roulette wheels and dice).

The coin, however, doesn't know anything about any mean or rubber band, and if it's landed heads 519 times and tails 481 times, the difference between its heads total and its tails total is just as likely to grow as to shrink. This is true despite the fact that the proportion of heads does approach ½ as the number of coin flips increases. (The gambler's fallacy should be distinguished from another phenomenon, regression to the mean, which is valid. If the coin is flipped a thousand more times, it is more likely than not that the number of heads on the second thousand flips would be smaller than 519.)

In terms of ratios, coins behave nicely: the ratio of heads to tails gets closer to 1 as the number of flips grows. In terms of absolute numbers, coins behave badly: the difference between the number of heads and the number of tails tends to get bigger as we continue to flip the coin, and the changes in lead from head to tail or vice versa tend to become increasingly rare.

If even fair coins behave so badly in an absolute sense, it's not surprising that some people come to be known as "losers" and others as "winners" though there is no real difference between them other than luck. Unfortunately perhaps, people are more sensitive to absolute differences between people than they are to rough equalities between them. If Peter and Paul have won, respectively, 519 and 481 trials, Peter will likely be termed a winner and Paul a loser. Winners (and losers) are often, I would guess, just people who get stuck on the right (or wrong) side of even. In the case of coins, it can take a long, long time for the lead to switch, longer often than the average life.

The surprising number of consecutive runs of heads or tails of various lengths give rise to further counter-intuitive notions. If Peter and Paul flip a fair coin every day to determine who pays for lunch, then it's more likely than not that at some time within about nine weeks Peter will have won five lunches in a row, as will have Paul. And at some period within about five to six years it's likely that each will have won ten lunches in a row.

Most people don't realize that random events generally can seem quite ordered. The following is a computer printout of a random sequence of Xs and Os, each with probability ½.

O X X X O O O X X X O X X X O X X X X O O X X O X X
O X O O X O X O O O O X O X X O O O X X X O X O X X

```
X X X X X X X O X X X O X O X X X X O X O O X X X O
O O X X X X X O O X X O O O X X O O O O O X X O O X
X X X X X O X X X X O O X X X X O O X X O X X O O X
X O X O X O O X X X O X X O X X X X O X X O X X X X
X X X X X O X X X X O O O O O X O O X X X O O X X
X X O O X O O X O X X X O X X X X O O O O X O X O X
X O X X X O O X X O O O O X X X X O O O O X X X X
O X X O O X X X X X X O X X O O O O O O O X O X X X
X X O O O X X O X X X O O O O X O X O X O O X X X X
O X O X X X O X X O O X X O X O O X O O X X X O X X
```

Note the number of runs and the way there seem to be clumps and patterns. If we felt compelled to account for these, we would have to invent explanations that would of necessity be false. Studies have been done, in fact, in which experts in a given field have analyzed such random phenomena and come up with cogent "explanations" for the patterns.

With this in mind, think of some of the pronouncements of stock analysts. The daily ups and downs of a particular stock, or of the stock market in general, certainly aren't completely random as the Xs and Os above are, but it's safe to say that there is a very large element of chance involved. You might never guess this, however, from the neat post hoc analyses that follow each market's close. Commentators always have a familiar cast of characters to which they can point to explain any rally or any decline. There's always profit-taking or the federal deficit or something or other to account for a bearish turn, and improved corporate earnings or interest rates or whatever to account for a bullish one. Almost never does a commentator say that the market's activity for the day or even the week was largely a result of random fluctuations.

The Hot Hand and the Clutch Hitter

The clumps, runs, and patterns that random sequences evince can to an extent be predicted. Sequences of heads and tails of a given length, say twenty flips, generally have a certain number of consecutive runs of heads. A sequence of twenty coin flips which resulted in ten heads followed by ten tails (HHHHHHHHHHHTTTTTTTTTT) is said to have just one run of heads. A sequence of twenty coin flips which resulted in heads and tails alternating (HTHTHTHTHTHTHTHTHTHT) is said to have ten runs of heads. Both these sequences are unlikely to be randomly generated. A sequence of twenty flips with six runs of heads (say, HHTHHTHTTHHHTTHHTTHT) is more likely to have been generated at random.

Criteria like this can be used to determine how likely it is that sequences of heads and tails or Xs and Os or hits and misses are randomly generated. In fact, psychologists Amos Tversky and Daniel Kahneman have analyzed the sequences of hits and misses of professional basketball players whose basket-making was about 50 percent and found that they seemed to be completely random —that a "hot hand" in basketball, one that would result in an inordinate number of long streaks (runs) of consecutive baskets, just didn't seem to exist. The streaks that did occur were most likely due to chance. If a player attempts twenty shots per night, for example, the probability is surprisingly almost 50 percent that he will hit at least four straight baskets sometime during the game. There's a 20 percent to 25 percent probability that he will achieve a streak of at least five straight baskets sometime during the game, and approximately a 10 percent chance that he will have a streak of six or more consecutive baskets.

Refinements can be made when the shooting percentage is other than 50 percent, and similar results seem to hold. A player who scores 65 percent of his shots, say, scores his points in the way a biased coin which lands heads 65 percent of the time "scores" its heads; i.e., each shot is independent of the last.

I've always suspected that notions like a "hot hand" or a "clutch hitter" or a "team that always comes back" were exaggerations used by sportswriters and sportscasters just to have something to talk about. There surely is something to these terms, but too often they're the result of minds intent on discovering meaning where there is only probability.

A very long hitting streak in baseball is a particularly amazing sort of record, so unlikely as to seem virtually unachievable and almost immune to probabilistic prediction. Several years ago Pete Rose set a National League record by hitting safely in forty-four consecutive games. If we assume for the sake of simplicity that he batted .300 (30 percent of the time he got a hit, 70 percent of the time he didn't) and that he came to bat four times per game, the chances of his not getting a hit in any given game were, assuming independence, $(.7)^4 = .24$. (Remember, independence means he got hits in the same way a coin which lands heads 30 percent of the time gets heads.) So the probability he would get at least one hit in any given game was $1 - .24 = .76$. Thus, the chances of his getting a hit in any given sequence of forty-four consecutive games were $(.76)^{44} = .0000057$, a tiny probability indeed.

The probability that he would hit in a consecutive string of exactly forty-four games at some time during the 162-game season is higher—.000041 (determined by adding up the ways in which he might hit safely in some string of exactly forty-four consecutive games, and ignoring the negligible probability of more than one streak). The probability that he'd hit safely in at least forty-four straight games is about four times higher still. If we multiply this latter figure by the number of players in the Major Leagues (adjusting the figure drastically downward for lower batting averages) and then multiply by the approximate number of years there has been baseball (adjusting for the various numbers of players in different years), we determine that it's actually not unlikely that some Major Leaguer should at some time have hit safely in at least forty-four consecutive games.

One last remark: I've examined Rose's streak of forty-four games rather than DiMaggio's seemingly more impressive streak of fifty-six games because, given the differences in their respective batting averages, Rose's streak was a slightly more unlikely accomplishment (even with Rose's longer season of 162 games).

Rare events such as batting streaks that are the result of chance are not individually predictable, yet the pattern of their occurrence is probabilistically describable. Consider a more prosaic kind of event. One thousand married couples who desire to have three children each are tracked for ten years, during which time 800 of them, assume, do manage to produce three children. The probability any given couple has three girls is $\frac{1}{2} \times \frac{1}{2} \times \frac{1}{2} = \frac{1}{8}$, so approximately a hundred of these 800 couples will have three girls each. By symmetry, about a hundred of the couples will have three boys each. There are three different sequences in which a family might have two girls and a boy—GGB, GBG, or BGG, where the order of the letters indicates birth order—and each of the three sequences has the same probability of $\frac{1}{8}$, or $(\frac{1}{2})^3$. Thus, the probability of having two girls and a boy is $\frac{3}{8}$, and so approximately 300 of the 800 couples will have such a family. By symmetry, about 300 couples will also have two boys and a girl.

Nothing is very surprising about the above, but the same sort of probabilistic description (utilizing mathematics slightly more difficult than the above binomial distribution) is possible with very rare events. The number of accidents each year at a certain intersection, the number of rainstorms per year in a given desert, the number of cases of leukemia in a specified county, the annual number of deaths due to horse kicks among certain cavalry units of the Prussian Army have all been described quite accurately by the so-called Poisson probability distribution. It's necessary first to know roughly how rare the event is. But if you do know, you can use this information along with the Poisson formula to get a quite accurate idea of, for example, in what percentage of years there would be no deaths due to horse kicks, in what percentage of years there would be one such death, in what percentage of years two, in what percentage three, and so on. Likewise you could predict the percentage of years in which there would be no desert rainstorms, one such storm, two storms, three, and so on.

In this sense, even very rare events are quite predictable.

3

Pseudoscience

When asked why he doesn't believe in astrology, the logician Raymond Smullyan responds that he's a Gemini, and Geminis never believe in astrology.

Sample of supermarket tabloid headlines: Miracle Pickup Truck Can Heal the Sick. Giant Bigfoot Attacks Village. Seven-Year-Old Gives Birth to Twins in Toy Store. Scientists on Verge of Creating Plant People. Incredible Swami Has Stood on One Leg since 1969.

Inspect every piece of pseudoscience and you will find a security blanket, a thumb to suck, a skirt to hold. What have we to offer in exchange? Uncertainty! Insecurity!
> —Isaac Asimov in the tenth-anniversary issue of
> *The Skeptical Inquirer*

To follow foolish precedents, and wink with both our eyes, is easier than to think.   —William Cowper

Innumeracy, Freud, and Pseudoscience

Innumeracy and pseudoscience are often associated, in part because of the ease with which mathematical certainty can be invoked to bludgeon the innumerate into a dumb acquiescence. Pure mathematics does indeed deal with certainties, but its applications are only as good as the underlying empirical assumptions, simplifications, and estimations that go into them.

Even such fundamental mathematical verities as "equals can be substituted for equals," or "1 and 1 are 2," can be misapplied: one cup of water plus one cup of popcorn are not equal to two cups of soggy popcorn, and "Infant Physician Duvalier" just doesn't have the same impact as "Baby Doc." Similarly, President Reagan may believe that Copenhagen is in Norway, but even though Copenhagen equals the capital of Denmark, it can't be concluded that Reagan believes the capital of Denmark is in Norway. In so-called intentional contexts like the above, the substitution doesn't always work.

If these basic principles can be misinterpreted, it shouldn't be surprising that more esoteric mathematics can be, too. If one's model or one's data are no good, the conclusions that follow won't be either. Applying old mathematics, in fact, is often more difficult than discovering new mathematics. Any bit of nonsense can be computerized—astrology, biorhythms, the *I Ching*—but that doesn't make the nonsense any more valid. Linear statistical projections, to cite a frequently abused model, are often invoked so thoughtlessly that it wouldn't be surprising to see someday that the projected waiting period for an abortion is one year.

This sort of careless reasoning is hardly limited to the uneducated. One of Freud's closest friends, a surgeon named Wilhelm Fliess, invented biorhythmic analysis, a practice based on the notion that various aspects of one's life follow rigid periodic cycles which begin at birth. Fliess pointed out to Freud that 23 and 28, the periods for some metaphysical male and female principles respectively, had the special property that if you add and subtract appropriate multiples of them, you can attain any number. Stated a little differently: any number at all can be expressed as $23X + 28Y$ for suitable choices of X and Y. For example, $6 = (23 \times 10) + (28 \times -8)$. Freud was so impressed with this that for years he was an ardent believer in biorhythms and thought that he would die at the age of fifty-one, the sum of 28 and 23. As it turns out, not only 23 and 28 but any two numbers that are relatively prime—that is, have no factors in common—have the property that any number can be expressed in terms of them. So even Freud suffered from innumeracy.

Freudian theory suffers from a more serious problem as well. Consider the statement: "Whatever God wills, happens." People may be able to take solace from it, but it's clear that the statement is not falsifiable and hence, as the English philosopher Karl Popper has insisted, not part of science. "Plane crashes always come in threes." You always hear that, too, and if you wait long enough, of course, everything comes in threes.

Popper has criticized Freudianism for claims and predictions which, though perhaps comforting or suggestive in one way or another, are, like the above statements, largely unfalsifiable. For example, an orthodox psychoanalyst might predict a certain kind of neurotic behavior. When the patient doesn't react in the predicted way, but in a very different manner, the analyst may attribute the opposite behavior to "reaction-formation." Likewise, when a Marxist predicts that the "ruling class" will act in an exploitive manner and instead something quite contrary takes place, he may attribute the outcome to an attempt by the ruling class to co-opt the "working class." There always seem to be escape clauses which can account for anything.

This is certainly not the place to argue whether or not Freudianism and Marxism should be deemed pseudosciences, but a tendency to confuse factual statements with empty logical formulations leads to sloppy thought. For example, the statements "UFOs contain extraterrestrial visitors" and "UFOs are unidentified flying objects" are two entirely different assertions. I once gave a lecture in which a listener thought that I subscribed to a belief in extraterrestrial visitors, when all I had said was that there undoubtedly were many cases of UFOs. A similar confusion is satirized by Molière when he has his pompous doctor announce that his sleeping potion works because of its dormitive virtue. Since mathematics is the quintessential way to make impressive-sounding claims which are devoid of factual content ("Scientists reveal that 36 inches equal 1 yard on the planet Pluto"), it's perhaps not surprising that it is an ingredient in a number of pseudosciences. Abstruse calculations, geometric forms and algebraic terms, unusual correlations—all have been used to adorn the silliest drivel.

Parapsychology

Interest in parapsychology is very old, yet the simple fact is that there have been no repeatable studies which have demonstrated its existence, Uri Geller and other charlatans notwithstanding. ESP (extrasensory perception) in particular has never been shown in any controlled experiment, and the few "successful" demonstrations have occurred in studies that were fatally flawed. Rather than rehash them, I'd like to make some general observations.

The first one, which is embarrassingly obvious, is that ESP runs afoul of the fundamental common-sense principle that the normal senses must somehow be involved for communication to take place. When confidential information leaks out of an organization, people suspect a spy, not a psychic. Hence, it is the presumption of common sense and science that these ESP phenomena don't exist, and the burden of proof is on those who maintain that there are such phenomena.

This raises probabilistic considerations. Because of the way ESP is defined—communication without any normal sensory mechanisms—there is no way to tell the difference between a single incidence of ESP and a chance guess. They look exactly the same, just as a particular correct answer on a true-false test looks the same whether the test taker is a straight-A student or someone who's guessing at every question. Since we can't ask the ESP subjects to justify their responses, as we can the true-false test takers, and since by definition there's no sensory mechanism into whose functioning we can inquire, the only way we can demonstrate the existence of ESP is by statistical test: run enough trials and see if the number of correct responses is sufficiently large so as to rule out chance as the explanation. If chance is ruled out and there are no other explanations, ESP will be demonstrated.

There is, of course, a tremendous will to believe which accounts for many of the faulty experiments (such as J. B. Rhine's) and much of the outright chicanery (such as S. G. Soal's) that seem to characterize the paranormal field. Another factor is what is sometimes referred to as the Jeane Dixon effect (after the self-described psychic Jeane Dixon), whereby the relatively few correct predictions are heralded and therefore widely remembered, while the much more numerous incorrect predictions are conveniently forgotten or deemphasized. Supermarket tabloids never provide an end-of-year list of false predictions by psychics, nor do the more upscale New Age periodicals, which, despite a veneer of sophistication, are just as fatuous.

People often take the abundance and prominence of reports on psychics and parapsychological matters to be a kind of evidence of their validity. Where there's so much smoke (hot air, actually), people reason, there must be fire. The nineteenth century's infatuation with phrenology—to continue with a somewhat different heady concern—demonstrates the flimsiness of this line of thought. Then as now, pseudoscientific beliefs were not limited to the uneducated, and the belief that various psychological and mental attributes were discernible by examining the bumps and contours of one's head was widespread. Many corporations required prospective employees to submit to phrenological examinations as a condition of employment, and many couples contemplating marriage sought the advice of phrenologists. Periodicals devoted to the subject appeared, and references to its doctrines pervaded popular literature. Renowned educator Horace Mann saw phrenology as "the guide to philosophy and the handmaiden of Christianity," and Horace Greeley of "Go West, young man" fame advocated phrenology tests for all railroad engineers.

Descending to more pedestrian matters, let's consider the

firewalker's practice of walking barefoot on hot wood coals. The practice has often been cited as an example of "mind over matter" and you don't have to be innumerate to be impressed initially with such a feat (or with such feet). What makes this phenomenon less remarkable is the relatively little-known fact that dehydrated wood has an extremely low heat content and very poor heat conductivity. Just as it's possible to put your hand into a hot oven without burning yourself as long as you don't touch the metal oven racks, a person can walk quickly across burning wood coals without any serious harm to his feet. Of course, quasi-religious talk about mind control is more appealing than a discussion of heat content and conductivity. This, combined with the fact that these walkings take place in the evening to heighten the contrast between the cool night air and surrounding darkness and the hot glowing coals, accounts for the dramatic impact of firewalking.

Many other examples of pseudoscience (auras, crystal power, pyramids, the Bermuda triangle, etc.) are unmasked in *The Skeptical Inquirer*, a delightful quarterly publication of CSICOP, the Committee for the Scientific Investigation of Claims of the Paranormal, published by philosopher Paul Kurtz in Buffalo, New York.

Predictive Dreams

Another presumed kind of extrasensory perception is the predictive dream. Everyone has an Aunt Matilda who had a vivid dream of a fiery car crash the night before Uncle Mortimer wrapped his Ford around a utility pole. I'm my own Aunt Matilda: when I was a kid I once dreamed of hitting a grand-slam home run and two days later I hit a bases-loaded triple. (Even believers in precognitive experiences don't expect an exact correspondence.) When one has such a dream and the predicted event happens, it's hard not to believe in precognition. But, as the following derivation shows, such experiences are more rationally accounted for by coincidence.

Assume the probability to be one out of 10,000 that a particular dream matches in a few vivid details some sequence of events in real life. This is a pretty unlikely occurrence, and means that the chances of a nonpredictive dream are an overwhelming 9,999 out of 10,000. Also assume that whether or not a dream matches experience one day is independent of whether or not some other dream matches experience some other day. Thus, the probability of having two successive nonmatching dreams is, by the multiplication principle for probability, the product of 9,999/10,000 and 9,999/10,000. Likewise, the probability of having N straight nights of nonmatching dreams is $(9,999/10,000)^N$; for a year's worth of nonmatching or nonpredictive dreams, the probability is $(9,999/10,000)^{365}$.

Since $(9,999/10,000)^{365}$ is about .964, we can conclude that about 96.4 percent of the people who dream every night will have only nonmatching dreams during a one-year span. But that means that about 3.6 percent of the people who dream every night will have a predictive dream. 3.6 percent is not such a small fraction; it translates into millions of apparently precognitive dreams every year. Even if we change the probability to one in a million for such a predictive dream, we'll still get huge numbers of them by chance alone in a country the size of the United States. There's no need to invoke any special parapsychological abilities; the ordinariness of apparently predictive dreams does not need any explaining. What would need explaining would be the nonoccurrence of such dreams.

The same could be said about a wide variety of other unlikely events and coincidences. Periodically, for example, there are reports of some incredible collection of coincidences linking two people, a phenomenon whose probability, let's say, is estimated to be one in a trillion (1 divided by 10^{12}, or 10^{-12}). Should we be impressed? Not necessarily.

Since by the multiplication principle there are $(2.5 \times 10^8 \times 2.5 \times 10^8)$ or 6.25×10^{16} different pairs of people in the United States, and since we're assuming the probability of this collection of coincidences to be about 10^{-12}, the average number of "incredible" linkages we can expect is 6.25×10^{16} times 10^{-12}, or about 60,000. It's not so surprising, then, that occasionally one of these 60,000 strange connections comes to light.

One collection of coincidences too unlikely to be dismissed in this way is provided by the case of the proverbial monkey accidentally typing out Shakespeare's *Hamlet*. The probability of this occurring is $(1/35)^N$ (where N is the number of symbols in *Hamlet*, maybe 200,000, and 35 is the number of typewriter symbols, including letters, punctuation symbols, and the blank space). This number is infinitesimal—zero, for all practical purposes. Though some have taken this tiny probability as an argument for "creation science," the only thing it clearly indicates is that monkeys seldom write great plays. If they want to, they shouldn't waste their time trying to peck one out accidentally but should instead evolve into something that has a better chance of writing *Hamlet*. Incidentally, why is the question never put as follows: What is the probability that Shakespeare, by randomly flexing his muscles, might accidentally have found himself swinging through the trees like a monkey?

Li'l Ol' Me and the Stars

Astrology is a particularly widespread pseudoscience. The shelves of bookstores are stuffed with books on the subject, and almost every newspaper publishes a daily horoscope. A 1986 Gallup poll reports that 52 percent of American teenagers believe in it, and a distressing number of people in all walks of life seem to accept at least some of its ancient claims. I say "distressing" because if people believe astrologers and astrology, it's frightening to consider whom or what else they'll believe. It's especially so when, like President Reagan, they have immense power to act on these beliefs.

Astrology maintains that the gravitational attraction of the planets at the time of one's birth somehow has an effect on one's personality. This seems very difficult to swallow, for two reasons: (*a*) no physical or neurophysiological mechanism through which this gravitational (or other sort of) attraction is supposed to act is ever even hinted at, much less explained; and (*b*) the gravitational pull of the delivering obstetrician far outweighs that of the planet or planets involved. Remember that the gravitational force an object exerts on a body—say, a newborn baby—is proportional to the object's mass but inversely proportional to the square of the distance of the object from the body—in this case, the baby. Does this mean that fat obstetricians deliver babies that have one set of personality characteristics, and skinny ones deliver babies that have quite different characteristics?

These deficiencies of astrological theory are less visible to the innumerate, who are not likely to concern themselves with mechanisms, and who are seldom interested in comparing magnitudes. Even without a comprehensible theoretical foundation, however, astrology would deserve respect if it worked, if there were some empirical support for the accuracy of its claims. But, alas, there is no correlation between the date of one's birth and scores on any standard personality test.

Experiments have been performed (recently, by Shawn Carlson at the University of California) in which astrologers have been given three anonymous personality profiles, one of which was the client's. The client supplied all the relevant astrological data about his life (via questionnaire, not face-to-face) and the astrologer was required to pick the personality profile of the client. There were 116 clients altogether, and they were presented to thirty top (as judged by their peers) European and American astrologers. The result: the astrologers picked the correct personality profile for the clients about one out of three times, or no better than chance.

John McGervey, a physicist at Case Western Reserve University, looked up the birth dates of more than 16,000 scientists listed in *American Men of Science* and 6,000 politicians listed in *Who's Who in American Politics* and found the distribution of their signs was random, the signs uniformly distributed throughout the year. Bernard Silverman at Michigan State University obtained the records of 3,000 married couples in Michigan and found no correlation between their signs and astrologers' predictions about compatible pairs of signs.

Why, then, do so many people believe in astrology? One obvious reason is that people read into the generally vague astrological pronouncements almost anything they want to, and thus invest them with a truth which is not inherent in the pronouncements themselves. They're also more likely to remember true "predictions," overvalue coincidences, and ignore everything else. Other reasons are its age (of course, ritual murder and sacrifice are as old), its simplicity in principle and comforting complexity in practice, and its flattering insistence on the connection between the starry vastness of the heavens and whether or not we'll fall in love this month.

One last reason, I would guess, is that during individual sessions astrologers pick up on clues about clients' personalities from their facial expressions, mannerisms, body language, etc. Consider the famous case of Clever Hans, the horse who seemed to be able to count. His trainer would roll a die and ask him what number appeared on the die's face. Hans would slowly paw the ground the appropriate number of times and then stop, much to the amazement of onlookers. What was not so noticeable, however, was that the trainer stood stone-still until the horse pawed the correct number of times, and then, consciously or not, stirred slightly, which caused Hans to stop. The horse was not the source of the answer but merely a reflection of the trainer's knowledge of the answer. People often unwittingly play the role of trainer to astrologers who, like Hans, reflect their clients' needs.

The best antidote to astrology in particular and to pseudoscience in general is, as Carl Sagan has written, real science, whose wonders are as amazing but have the added virtue of probably being real. After all, it's not the outlandishness of its conclusions that makes something a pseudoscience: lucky guesses, serendipity, bizarre hypotheses, and even an initial gullibility all play a role in science as well. Where pseudosciences fail is in not subjecting their conclusions to a test, in not linking them in a coherent way to other statements which have withstood scrutiny. It's hard for me to imagine Shirley MacLaine, for example, rejecting the reality of some seemingly paranormal event such as trance channeling merely because there isn't enough evidence for it, or because there is a better alternative explanation.

Extraterrestrial Life, Yes; Visitors in UFOs, No

In addition to astrology, innumerates are considerably more likely than others to believe in visitors from outer space. Whether or not there have been such visits is a question distinct from whether or not there is other conscious life in the universe. I'll develop some very approximate estimates to indicate why, though there probably are other life forms in our very galaxy, they most likely haven't paid us a courtesy call (despite the claims of books such as Budd Hopkins's *The Intruders* and Whitley Strieber's *Communion*). The estimates provide a good example of how numerical horse sense can check pseudoscientific ravings.

If intelligence developed naturally on earth, it is difficult to see why the same process wouldn't have occurred elsewhere. What's needed is a system of physical elements capable of many different combinations, and a source of energy through the system. The energy flux causes the system to "explore" various combinations of possibilities, until a small collection of stable, complex, energy-storing molecules develops, followed by the chemical evolution of more complex compounds, including some amino acids, from which proteins are constructed. Eventually, primitive life develops, and then shopping malls.

It's estimated that there are approximately 100 billion stars (10^{11}) in our galaxy, of which, say, $1/10$th support a planet. Of these approximately 10 billion stars, perhaps one out of a hundred contains a planet which lies within the life zone of the star, not too close for its solvent, water or methane or whatever, to boil away and not too far to be frozen solid. Now we're down to approximately 100 million stars (10^8) in our galaxy which could support life. Since most of them are considerably smaller than our sun, only about ⅒th of these stars should be considered reasonable candidates for supporting planets with life. Still, this leaves us with 10 million stars (10^7) in our galaxy alone capable of supporting life, of which perhaps ⅒th have already developed life! Let's assume that there are indeed 10^6—or a million—stars with planets which support life in our own galaxy. Why don't we see any evidence?

One reason is that our galaxy is a big place, having a volume of about 10^{14} cubic light-years where a light-year is the distance light travels in one year at 186,000 miles per second—about 6 trillion miles. Thus, on average, each of these million stars has 10^{14} divided by 10^6 cubic light-years of volume for itself; that's 10^8 cubic light-years of volume for each star assumed to support life. The cube root of 10^8 is approximately 500, meaning that the average distance between any one of the galaxy's life-supporting stars and its closest neighbor would be 500 light-years—about ten billion times the distance between the earth and the moon! The distance between close "neighbors," even if it were considerably less than the average, would seem to preclude frequent popping in for a chat.

The second reason we would be quite unlikely to see any little green men is that possible civilizations are bound to be scattered in time, coming into existence and then dying out. In fact, it could well be the case that life, once it becomes complex, is inherently unstable and will self-destruct within a few thousand years. Even if such advanced life forms had an average duration of 100 million years (the time from early mammals to a possible twentieth-century nuclear holocaust), these life forms spread uniformly over the 12–15 billion-year history of our galaxy would result in fewer than 10,000 stars in our galaxy supporting advanced life at any one time. The average distance between neighbors would jump to more than 2,000 light-years.

The third reason we haven't had any tourists is that even if life has developed on a number of planets within our galaxy, there's probably little likelihood they'd be interested in us. The life forms could be large clouds of methane gas, or self-directed magnetic fields, or large plains of potato-like beings, or giant planet-sized entities which spend their time singing complex symphonies, or more likely a sort of planetary scum adhering to the sides of rocks facing their sun. There's little reason to suppose that any of the above would share our goals or psychology and try to reach us.

In short, though there probably is life on other planets in our galaxy, the sightings of UFOs are almost certainly just that—sightings of unidentified flying objects. Unidentified, but not unidentifiable or alien.

Fraudulent Medical Treatments

Medicine is a fertile area for pseudoscientific claims for a simple reason. Most diseases or conditions (*a*) improve by themselves; (*b*) are self-limiting; or (*c*) even if fatal, seldom follow a strictly downward spiral. In each case, intervention, no matter how worthless, can appear to be quite efficacious.

This becomes clearer if you assume the point of view of a knowing practitioner of fraudulent medicine. To take advantage of the natural ups and downs of any disease (as well as of any placebo effect), it's best to begin your worthless treatment when the patient is getting worse. In this way, anything that happens can more easily be attributed to your wonderful and probably expensive intervention. If the patient improves, you take credit; if he remains stable, your treatment stopped his downward course. On the other hand, if the patient worsens, the dosage or intensity of the treatment was not great enough; if he dies, he delayed too long in coming to you.

In any case, the few instances in which your intervention is successful will likely be remembered (not so few, if the disease in question is self-limiting), while the vast majority of failures will be forgotten and buried. Chance provides more than enough variation to account for the sprinkling of successes that will occur with almost any treatment; indeed, it would be a miracle if there weren't any "miracle cures."

Much of the above applies as well to faith healers, psychic surgeons, and an assorted variety of other practitioners from homeopathic physicians to TV evangelists. Their prominence constitutes a strong argument for an infusion of healthy skepticism into our schools, a state of mind generally incompatible with innumeracy. (By this dismissive attitude toward these charlatans, however, I don't mean to advocate a rigid and dogmatic scientism or some kind of simpleminded atheism. There's a long way from Adonai to I Don't Know to I Deny—to adapt a line from poet Howard Nemerov—and plenty of room in the middle for reasonable people to feel comfortable.)

Even in outlandish cases, it's often difficult to refute conclusively some proposed cure or procedure. Consider a quack diet doctor who directs his patients to consume two whole pizzas, four birch beers, and two pieces of cheesecake for every breakfast, lunch, and dinner, and an entire box of fig bars with a quart of milk for a bedtime snack, claiming that other people have lost six pounds a week on such a regimen. When several patients follow his instructions for three weeks, they find they've gained about seven pounds each. Have the doctor's claims been refuted? Not necessarily, since he might respond that a whole host of auxiliary understandings weren't met: the pizzas had too much sauce, or the dieters slept sixteen hours a day, or the birch beer wasn't the right brand. The point is that one can usually find loopholes which will enable one to hold on to whatever pet theory one fancies.

The philosopher Willard Van Orman Quine goes even further and maintains that experience never forces one to reject any particular belief. He views science as an integrated web of interconnecting hypotheses, procedures, and formalisms, and argues that any impact of the world on the web can be distributed in many different ways. If we're willing to make drastic enough changes in the rest of the web of our beliefs, the argument goes, we can hold to our belief in the efficacy of the above diet, or indeed in the validity of any pseudoscience.

Less controversial is the contention that there are no clear-cut, easy algorithms that allow us to distinguish science from pseudoscience in all cases. The boundary between them is too fuzzy. Our unifying topics, number and probability, do, however, provide the basis for statistics, which, together with logic, constitutes the foundation of the scientific method, which will eventually sort matters out if anything can. However, just as the existence of pink does not undermine the distinction between red and white, and dawn doesn't indicate that day and night are really the same, this problematic fringe area, Quinian arguments notwithstanding, doesn't negate the fundamental differences between science and its impostors.

Conditional Probability, Blackjack, and Drug Testing

One needn't be a believer in any of the standard pseudosciences to make faulty claims and invalid inferences. Many mundane mistakes in reasoning can be traced to a shaky grasp of the notion of conditional probability. Unless the events A and B are independent, the probability of A is different from the probability of A given that B has occurred. What does this mean?

To cite a simple example, the probability that a person chosen at random from the phone book is over 250 pounds is quite small. However, if it's known somehow that the person chosen is over six feet four inches tall, then the conditional probability that he or she also weighs more than 250 pounds is considerably higher. The probability of rolling a pair of dice and getting a 12 is $\frac{1}{36}$. The conditional probability of getting a 12 when you know you have gotten at least an 11 is $\frac{1}{3}$. (The outcomes could only be 6,6; 6,5; 5,6 and thus there's one chance in three that the sum is 12, given that it's at least 11.)

A confusion between the probability of A given B and the probability of B given A is also quite common. A simple example: the conditional probability of having chosen a king card when it's known that the card is a face card—a king, queen, or jack—is $\frac{1}{3}$. However, the conditional probability that the card is a face card given that it's a king is 1, or 100 percent. The conditional probability that someone is an American citizen, given that he or she speaks English, is, let's assume, about ⅕. The conditional probability that someone speaks English, given that he or she is an American citizen, on the other hand, is probably about ¹⁹⁄₂₀.

Consider now some randomly selected family of four which is known to have at least one daughter. Say Myrtle is her name. Given this, what is the conditional probability that Myrtle's sibling is a brother? Given that Myrtle has a younger sibling, what is the conditional possibility that her sibling is a brother? The answers are, respectively, ⅔ and ½.

In general, there are four equally likely possibilities for a family with two children—BB, BG, GB, GG, where the order of the letters B (boy) and G (girl) indicates birth order. In the first case, the possibility BB is ruled out since Myrtle is a girl, and in two of the three other equally likely possibilities, there is a boy, Myrtle's brother. In the second case, the possibilities BB and BG are ruled out since Myrtle, a girl, is the older sibling, and in one of the remaining two equally likely possibilities, there is a boy, Myrtle's brother. In the second case, we know more, accounting for the differing conditional probabilities.

Before I get to a serious application, I'd like to mention another con game which works because of confusion about conditional probability. Imagine a man with three cards. One is black on both sides, one red on both sides, and one black on one side and red on the other. He drops the cards into a hat and asks you to pick one, but only to look at one side; let's assume it's red. The man notes that the card you picked couldn't possibly be the card that was black on both sides, and therefore it must be one of the other two cards—the red-red card or the red-black card. He offers to bet you even money that it is the red-red card. Is this a fair bet?

At first glance, it seems so. There are two cards it could be; he's betting on one, and you're betting on the other. But the rub is that there are two ways he can win and only one way you can win. The visible side of the card you picked could be the red side of the red-black card, in which case you win, or it could be one side of the red-red card, in which case he wins, or it could be the other side of the red-red card, in which case he also wins. His chances of winning are thus ⅔. The conditional probability of the card being red-red given that it's not black-black is ½, but that's not the situation here. We know more than just that the card is not black-black; we also know a red side is showing.

Conditional probability also explains why blackjack is the only casino game of chance in which it makes sense to keep track of past occurrences. In roulette, what's occurred previously has no effect on the probability of future spins of the wheel. The probability of red on the next spin is 18/38, the same as the conditional probability of red on the next spin given that there have been five consecutive reds. Likewise with dice: the probability of rolling a 7 with a pair of dice is 1/6, the same as the conditional probability of rolling a 7 given that the three previous rolls have been 7s. Each trial is independent of the past.

A game of blackjack, on the other hand, is sensitive to its past. The probability of drawing two aces in succession from a deck of cards is not (4/52 × 4/52) but rather (4/52 × 3/51), the latter factor being the conditional probability of choosing another ace given that the first card chosen was an ace. Likewise, the conditional probability that a card drawn from a deck will be a face card, given that only two of the thirty cards drawn so far have been face cards, is not 12/52 but a much higher 10/22. This fact—that (conditional) probabilities change according to the composition of the remaining portion of the deck—is the basis for various counting strategies in blackjack that involve keeping track of how many cards of each type have already been drawn and increasing one's bet when the odds are (occasionally and slightly) in one's favor.

I've made money at Atlantic City using these counting strategies, and even considered having a specially designed ring made which would enable me to count more easily. I decided against it, though, since, unless one has a large bankroll, the rate at which one wins money is too slow to be worth the time and intense concentration required.

An interesting elaboration on the concept of conditional probability is known as Bayes' theorem, first proved by Thomas Bayes in the eighteenth century. It's the basis for the following rather unexpected result, which has important implications for drug or AIDS testing.

Assume that there is a test for cancer which is 98 percent accurate; i.e., if someone has cancer, the test will be positive 98 percent of the time, and if one doesn't have it, the test will be negative 98 percent of the time. Assume further that .5 percent—one out of two hundred people—actually have cancer. Now imagine that you've taken the test and that your doctor somberly informs you that you've tested positive. The question is: How depressed should you be? The surprising answer is that you should be cautiously optimistic. To find out why, let's look at the conditional probability of your having cancer, given that you've tested positive.

Imagine that 10,000 tests for cancer are administered. Of these, how many are positive? On the average, 50 of these 10,000 people (.5 percent of 10,000) will have cancer, and so, since 98 percent of them will test positive, we will have 49 positive tests. Of the 9,950 cancerless people, 2 percent of them will test positive, for a total of 199 positive tests (.02 × 9,950 = 199). Thus, of the total of 248 positive tests (199 + 49 = 248), most (199) are false positives, and so the conditional probability of having cancer given that one tests positive is only 49/248, or about 20 percent! (This relatively low percentage is to be contrasted with the conditional probability that one tests positive, given that one has cancer, which by assumption is 98 percent.)

This unexpected figure for a test that we assumed to be 98 percent accurate should give legislators pause when they contemplate instituting mandatory or widespread testing for drugs or AIDS or whatever. Many tests are less reliable: a recent article in *The Wall Street Journal*, for example, suggests that the well-known Pap test for cervical cancer is only 75 percent accurate.

Lie-detection tests are notoriously inaccurate, and calculations similar to the above demonstrate why truthful people who flunk polygraph tests usually outnumber liars. To subject people who test positive to stigmas, especially when most of them may be false positives, is counterproductive and wrong.

Numerology

Less worrisome than inaccurate tests is numerology, the last pseudoscience I want to discuss, and my favorite. It is a very old practice common to a number of ancient and medieval societies and involves the assignment of numerical values to letters and the consequent reading of significance into the numerical equality between various words and phrases.

The numerical values of the letters in the Hebrew word for "love" (*ahavah*) add up to 13, the same total as the letters in the word for "one" (*ehad*). Since "one" is short for "one God," the equality of the two words was deemed significant by many, as was the fact that their sum is 26, the numerical equivalent of "Yahweh," the divine name of God.

The number 26 was important for other reasons: in verse 26 of the first chapter of Genesis, God says: "Let us make man in our image"; Adam and Moses were separated by 26 generations; and the difference between the numerical equivalent of Adam (45) and that of Eve (19) is 26.

The rabbis and cabalists who engaged in numerology (gematria) used a variety of other systems as well, sometimes disregarding powers of 10—taking 10 to be 1, 20 to be 2, and so on. Thus, since the first letter of "Yahweh" was assigned a value of 10, it could, when the occasion demanded, be assigned a value of 1, making "Yahweh" equal in value to 17, the same as the numerical equivalent of the word for "good" (*tov*). At other times they considered the squares of the numerical values of the letters, in which case "Yahweh" would equal 186, the same as the word for "place" (*Maqom*), another way of referring to God.

The Greeks, too, engaged in numerological practice (isopsephia), both in antiquity, with the number mysticism of Pythagoras and his school, and especially later, with the introduction of Christianity. In this system the Greek word for "God" (*Theos*) had a numerical value of 284, as did the words for "holy" and "good." The numerical value of the letters alpha and omega, the beginning and the end, was 801, the same as the word for "dove" (*peristera*), and was supposed to be a mystical corroboration of the Christian belief in a Trinity. The Greek Gnostics noted that the Greek word for "Nile River" had a numerical value of 365, indicating the annual nature of its floods.

Christian mystics devoted much energy to deciphering the number 666, said by John the Apostle to designate the name of the Beast of the Apocalypse, the Antichrist. The method used to assign numbers to letters was not specified, however, and so it's not entirely clear to whom the number refers. "Caesar Nero," the name of the first Roman Emperor to persecute the Christians, had a value of 666 in the Hebrew system, as did the word for "Latins" in the Greek system. The number has often been used in the service of ideology: a Catholic writer of the sixteenth century wrote a book whose gist was that Martin Luther was the Antichrist, since in the Latin system his name had a value of 666. Soon enough, some of Luther's followers responded that the words in the papal crown, "Vicar of the Son of God," had the value 666 if one added the Roman numerals corresponding to letters appearing in the phrase. More recently, the extreme fundamentalist right has noted that each word in the name Ronald Wilson Reagan has six letters.

Similar examples could be given of Moslem numerological practices. Such numerical readings (Jewish, Greek, Christian, and Moslem) were used not just to provide mystical confirmation of religious doctrine but also in soothsaying, dream interpretation, divination by numbers, etc. Often, they were opposed by orthodox clergy, but were very popular among the laity.

Even today, some of these numerological superstitions are not dead. I wrote a review for *The New York Times* of Georges Ifrah's *From One to Zero* (from which most of the above is taken) and referred in a completely neutral manner to the number 666, Martin Luther, and the papal crown. In response I received a half dozen deranged, anti-Semitic letters, some calling me the Antichrist. Procter and Gamble had similar but more severe problems a few years ago with reference to the numerico-symbolic nature of its logo.

Numerology, especially in its soothsaying and divinatory aspects, is in many ways a typical pseudoscience. It makes predictions and claims that are almost impossible to falsify since an alternative formulation consistent with what happened is always easy to dream up. Based on number, it has a limitless complexity to engage the ingenuity and creativity of its adherents, without burdening them with the need for validation or testing. Its expressions of equality are generally used to corroborate some existing doctrine, and little if any effort is expended to construct counter-examples. Surely, "God" must be numerically equivalent to phrases which deny doctrine, or to words which are sacrilegious or funny. (I'll forgo giving my examples.) Like many other pseudosciences, numerology is ancient, and acquires some respectability from its religious associations.

Still, if one subtracts all the superstitious elements from the subject, there's something appealing about the small residue that remains. Its purity (just numbers and letters) and tabula-rasa quality (like a Rorschach test) allow one maximum scope for seeing what one wants to see, for connecting what one wants to connect, for providing at the very least a limitless source of mnemonic devices.

Logic and Pseudoscience

Since numbers and logic are inextricably intertwined both theoretically and in the popular mind, it's perhaps not stretching matters too far to describe faulty logic as a kind of innumeracy. This assumption has in fact been implicit throughout much of this chapter. Let me end, then, with a couple of additional bad inferences which are further suggestive of the role that innumeracy—in the guise of fallacious logic—plays in pseudoscience.

Confusing a conditional statement—if A, then B—with its converse—if B, then A—is a very common mistake. A slightly unusual version of it occurs when people reason that if X cures Y, then lack of X must cause Y. If the drug dopamine, e.g., brings about a decrease in the tremors of Parkinson's disease, then lack of dopamine must cause tremors. If some other drug relieves the symptoms of schizophrenia, then an excess of it must cause schizophrenia. One is not as likely to make this mistake when the situation is more familiar. Not too many people believe that since aspirin cures headaches, lack of aspirin in the bloodstream must cause them.

From a jar of fleas before him, the celebrated experimenter Van Dumholtz carefully removes a single flea, gently pulls off its back legs, and in a loud voice commands it to jump. He notes that it doesn't move and tries the same thing with a different flea. When he's finished, he compiles statistics and concludes confidently that a flea's ears are in its back legs. Absurd perhaps, but variants of this explanation in less transparent contexts might carry considerable force for people with strong enough preconceptions. Is the explanation any more absurd than that accepted by those who believe the woman who maintains that she is the channel through which a 35,000-year-old man expresses himself? Is it more strained than claims that the skepticism of onlookers systematically prevents the occurrence of certain paranormal phenomena?

What's wrong with the following not quite impeccable logic? We know that 36 inches = 1 yard. Therefore, 9 inches = ¼ of a yard. Since the square root of 9 is 3 and the square root of ¼ is ½, we conclude that 3 inches = ½ yard!

Disproving a claim that something exists is often quite difficult, and this difficulty is often mistaken for evidence that the claim is true. Pat Robertson, the former television evangelist and Presidential candidate, maintained recently that he couldn't prove that there weren't Soviet missile sites in Cuba and therefore there might be. He's right, of course, but neither can I prove that Big Foot doesn't own a small plot of land outside Havana. New Agers make all sorts of existence assertions: that ESP exists, that there have been instances of spoon bending, that spirits abound, that there are aliens among us, etc. Presented as I periodically am with these and other fantastical claims, I sometimes feel a little like a formally dressed teetotaler at a drunken orgy for reiterating that not being able to conclusively refute the claims does not constitute evidence for them.

Many more vignettes illustrating this and other simple logical errors might be cited, but the point is clear enough: both innumeracy and defective logic provide a fertile soil for the growth of pseudoscience. Why both are so widespread is the topic of the next chapter.

4

Whence Innumeracy?

Recent personal experience at a suburban fast-food restaurant: My order of a hamburger, French fries, and a Coke comes to $2.01, and the cashier, who's worked there for months at least, fumbles with the 6 percent tax chart at the side of the cash register, searching for the line that says $2.01—$.12. Accommodating their innumerate help, the larger franchises now have cash registers which have pictures, on the keys, of the items ordered and automatically add on the appropriate tax.

A study indicates that whether or not a department has a mathematics or a statistics requirement is the most important single determinant of where a woman will attend graduate school to study political science.

When I heard the learn'd astronomer where he lectured with much applause in the lecture-room / How soon unaccountable I became tired and sick.   —Walt Whitman

Remembrance of Innumeracies Past

Why is innumeracy so widespread even among otherwise educated people? The reasons, to be a little simplistic, are poor education, psychological blocks, and romantic misconceptions about the nature of mathematics. My own case was the exception that proves the rule. The earliest memory I have of wanting to be a mathematician was at age ten, when I calculated that a certain relief pitcher for the then Milwaukee Braves had an earned run average (ERA) of 135. (For baseball fans: He allowed five runs to score and retired only one batter.) Impressed by this extraordinarily bad ERA, I diffidently informed my teacher, who told me to explain the fact to my class. Being quite shy, I did so with a quavering voice and a reddened face. When I finished, he announced that I was all wrong and that I should sit down. ERAs, he asserted authoritatively, could never be higher than 27.

At the end of the season, *The Milwaukee Journal* published the averages of all Major League players, and since this pitcher hadn't played again, his ERA was 135, as I had calculated. I remember thinking of mathematics as a kind of omnipotent protector. You could prove things to people and they would have to believe you whether they liked you or not. So, still smarting from my perceived humiliation, I brought in the paper to show the teacher. He gave me a dirty look and again told me to sit down. His idea of a good education apparently was to make sure everyone remained seated.

Though not dominated by martinets like my teacher, early mathematics education is generally poor. Elementary schools by and large do manage to teach the basic algorithms for multiplication and division, addition and subtraction, as well as methods for handling fractions, decimals, and percentages. Unfortunately, they don't do as effective a job in teaching when to add or subtract, when to multiply or divide, or how to convert from fractions to decimals or percentages. Seldom are arithmetic problems integrated into other schoolwork—how much, how far, how old, how many. Older students fear word problems in part because they have not been asked to find solutions to such quantitative questions at the elementary level.

Although few students get past elementary school without

knowing their arithmetic tables, many do pass through without understanding that if one drives at 35 m.p.h. for four hours, one will have driven 140 miles; that if peanuts cost 40 cents an ounce and a bag of them costs $2.20, then there are 5.5 ounces of peanuts in the bag; that if ¼ of the world's population is Chinese and ⅕ of the remainder is Indian, then ³⁄₂₀ or 15 percent of the world is Indian. This sort of understanding is, of course, not the same as simply knowing that $35 \times 4 = 140$; that $(2.2)/(.4) = 5.5$; that $⅕ \times (1 - ¼) = ³⁄₂₀ = .15 = 15$ percent. And since it doesn't come naturally to many elementary students, it must be furthered by doing numerous problems, some practical, some more fanciful.

Estimation is generally not taught either, aside from a few lessons on rounding off numbers. The connection is rarely made that rounding off and making reasonable estimates have something to do with real life. Grade-school students aren't invited to estimate the number of bricks in the side of a school wall, or how fast the class speedster runs, or the percentage of students with bald fathers, or the ratio of one's head's circumference to one's height, or how many nickels are necessary to make a tower equal in height to the Empire State Building, or whether all those nickels would fit in their classroom.

Almost never is inductive reasoning taught or are mathematical phenomena studied with an eye toward guessing the relevant properties and rules. A discussion of informal logic is as common in elementary mathematics courses as is a discussion of Icelandic sagas. Puzzles, games, and riddles aren't discussed— in many cases, I'm convinced, because it's too easy for bright ten-year-olds to best their teachers. The intimate relationship between mathematics and such games has been explored most engagingly by mathematics writer Martin Gardner, whose many charming books and *Scientific American* columns would make exciting outside reading for high school or college students (were they but assigned), as might mathematician George Polya's *How to Solve It* or *Mathematics and Plausible Reading*. A delightful book with something of the flavor of these others, but at an elementary level, is *I Hate Mathematics* by Marilyn Burns. It's full of what elementary math textbooks rarely have—heuristic tips on problem solving and whimsy.

Instead, too many textbooks still list names and terms, with few if any illustrations. They note, for example, that addition is said to be an associative operation since (a + b) + c = a + (b + c). Seldom is any mention made of an operation which is non-associative, so the definition seems unnecessary at best. In any case, what can you do with this piece of information? Other terms seem to be introduced with no rationale other than that they look impressive when printed in boldface type inside a little box in the middle of the page. They satisfy many people's conception of knowledge as a kind of general botany where there's a place for everything and everything has its place. Mathematics as a useful tool or as a way of thinking or as a source of pleasure is a notion quite foreign to most elementary-education curricula (even to those whose textbooks are adequate).

One would think that, at this level, computer software would be available to help communicate the basics of arithmetic and its applications (word problems, estimation, etc.). Unfortunately, the programs we have at present are too often transcriptions onto television monitors of unimaginative lists of routine exercises taken from the textbooks. I'm not aware of any software which offers an integrated, coherent, and effective approach to arithmetic and its problem-solving applications.

Some of the blame for the generally poor instruction in elementary schools must ultimately lie with teachers who aren't sufficiently capable, and who too often have little interest in or appreciation of mathematics. In turn, some of the blame for that lies, I think, with schools of education in colleges and universities which place little or no emphasis on mathematics in their teacher-training courses. It's been my experience that students in secondary math education (as opposed to math majors) are generally among the worst students in my classes. The background in math of prospective elementary-school teachers is even worse; in many cases, nonexistent.

A partial solution might be the hiring of one or two mathematics specialists per elementary school who would move from room to room throughout the school day, supplementing (or teaching) the mathematics curriculum. I sometimes think it would be a good idea if math professors and elementary-school teachers switched places for a few weeks each year. No harm would come to the math majors and graduate students at the hands of the elementary-school teachers (in fact, the latter might learn something from the former), while the third-, fourth-, and fifth-graders might greatly benefit from exposure to mathematical puzzles and games competently presented.

A little digression. This connection between puzzles and mathematics persists through to graduate and research-level mathematics, and the same may be said of humor. In my book *Mathematics and Humor* I tried to show that both activities are forms of intellectual play, which often find common ground in brainteasers, puzzles, games, and paradoxes.

Both mathematics and humor are combinatorial, taking apart and putting together ideas for the fun of it—juxtaposing, generalizing, iterating, reversing (AIXELSYD). What if I relax this condition and strengthen that one? What does this idea—say, the knotting of braids—have in common with that one in some other seemingly disparate area—say, the symmetries of some geometric figure? Of course, this aspect of mathematics isn't very well known even to the numerate, since it's necessary to have some mathematical ideas first before you can play around with them. As well, ingenuity, a feeling for incongruity, and a sense of economical expression are crucial to both mathematics and humor.

Mathematicians, it may be noted, have a characteristic sense of humor which may be a result of their training. They have a tendency to take expressions literally, and this literal interpretation is often incongruous with the standard one and therefore comical. (Which two sports have face-offs? Ice hockey and leper boxing.) They indulge as well in reductio ad absurdum, the logical practice of taking any premise to an extreme, and in various sorts of combinatorial word play.

If mathematics education communicated this playful aspect
of the subject, formally at the elementary, secondary, or college
level or informally via popular books, I don't think innumeracy
would be as widespread as it is.

Secondary, College, and Graduate Education

Once students reach high school, the problem of teacher
competence becomes more critical. So many of the limited pool
of mathematically talented people now work in the computer
industry or in investment banking or related fields that I think
only substantial salary bonuses for well-qualified secondary-
school math teachers will keep the situation in our high schools
from getting worse. Since at this level a long list of education
courses is not as essential as having a mastery of the relevant
mathematics, certifying retired engineers and other science
professionals to teach mathematics might be of considerable
help. As it is, the basic elements of mathematical culture are in
many cases not being communicated to our students. Vieta in
1579 began to use algebraic variables—X, Y, Z, etc.—to sym-
bolize unknown quantities. A simple idea this, yet many high
school students today can't follow this four-hundred-year-old
method of reasoning: Let X be the unknown quantity, find an
equation which X satisfies, and then solve it in order to find the
value of the unknown.

Even when the unknowns are appropriately symbolized and
the relevant equation can be set up, the manipulations necessary
to solve it are too often only hazily understood. I wish I had five
dollars for every student who got through his or her high school
algebra class only to write, on a test in freshman calculus, that
$(X + Y)^2 = X^2 + Y^2$.

Approximately fifty years after Vieta's use of algebraic variables, Descartes devised a way of associating points on a plane
with ordered pairs of real numbers and, via this association, a
way of identifying algebraic equations with geometric curves.
The subject that grew out of this critical insight, analytic geometry, is essential to understanding calculus; yet our students

are coming out of high school unable to graph straight lines or parabolas.

Even the 2,500-year-old Greek idea of an axiomatic geometry—a few self-evident axioms being assumed, and from them the theorems being derived by logic alone—is not being effectively taught in secondary school. One of the most commonly used books in high school geometry classes makes use of more than a hundred axioms to prove a similar number of theorems! With so many axioms, all the theorems are surface ones requiring only three or four steps to prove; none has any depth.

In addition to some understanding of algebra, geometry, and analytic geometry, high school students should be exposed to some of the most important ideas of so-called finite mathematics. Combinatorics (which studies various ways of counting the permutations and combinations of objects), graph theory (which studies networks of lines and vertices and the phenomena which can be modeled by such), game theory (the mathematical analysis of games of all sorts), and especially probability, are increasingly important. In fact, the move to teach calculus in some high schools seems to me wrongheaded if it leads to the exclusion of the above topics in finite mathematics. (I'm writing here of an ideal high school curriculum. As the recent "Mathematics Report Card" administered by the Educational Testing Service has indicated, the majority of our high school students can barely solve the elementary problems I mentioned a few pages back.)

High school is the time to reach students. After they get to college, it's often too late for many of them who lack adequate backgrounds in algebra and analytic geometry. Even students who have a reasonable math background are not always aware of the extent to which other subjects are becoming "mathematicized," and they, too, take a minimum of mathematics in college.

Women, in particular, may end up in lower-paying fields because they do everything in their power to avoid a chemistry or an economics course with mathematics or statistics prerequisites. I've seen too many bright women go into sociology and too many dull men go into business, the only difference between them being that the men managed to scrape through a couple of college math courses.

The students who do major in mathematics in college, taking the basic courses in differential equations, advanced calculus, abstract algebra, linear algebra, topology, logic, probability and statistics, real and complex analysis, etc., have a large number of options, not only in mathematics and computer science but in an increasing variety of fields which utilize mathematics. Even when companies recruit for jobs that have nothing to do with mathematics, they often encourage math majors to apply, since they know that analytical skills will serve anyone well, whatever the job.

Mathematics majors who continue their studies will find that graduate education in mathematics, in great contrast to that at lower levels, is the best in the world. Unfortunately, by this time it's too late for most, and this preeminence in research doesn't filter down to lower levels, due in good measure to the failure of American mathematicians to reach an audience wider than the small number of specialists who read their research papers.

Excluding certain textbook authors, only a handful of mathematics writers have a lay audience of more than a thousand. Given this sorry fact, it's perhaps not surprising that few educated people will admit to being completely unacquainted with the names Shakespeare, Dante, or Goethe, yet most will openly confess their ignorance of Gauss, Euler, or Laplace, in some sense their mathematical analogues. (Newton doesn't count, since he's much more famous for his contributions to physics than for his invention of calculus.)

Even at the graduate and research level, there are ominous signs. So many foreign students come to do their graduate work here, and so few American students major in mathematics, that in many departments American graduate students are a minority. In fact, of 739 doctorates in mathematics awarded by American universities in 1986–87, slightly less than half, only 362, were conferred on United States citizens.

If mathematics is important (and it certainly is), then so is mathematics education. Mathematicians who don't deign to communicate their subject to a wider audience are a little like multimillionaires who don't contribute anything to charity. Given the relatively low salaries of many mathematicians, both failings might be overcome if multimillionaires supported mathematicians who wrote for a popular audience. (Just a thought.)

One argument mathematicians cite for not writing for a larger audience is the esoteric nature of their work. There's something to this, of course, but Martin Gardner, Douglas Hofstadter, and Raymond Smullyan are three obvious counterexamples. In fact, some of the ideas discussed in this book are quite sophisticated, yet the mathematical prerequisites for understanding them are truly minimal: some facility with arithmetic and an understanding of fractions, decimals, and percentages. It is almost always possible to present an intellectually honest and engaging account of any field, using a minimum of technical apparatus. This is seldom done, however, since most priesthoods (mathematicians included) are inclined to hide behind a wall of mystery and to commune only with their fellow priests.

In short, there is an obvious connection between innumeracy and the poor mathematical education received by so many people. Hence this jeremiad. Still, it's not the whole story, since there are many quite numerate people who have had little formal schooling. More debilitating mathematically than ineffective or insufficient education are psychological factors.

Innumeracy and the Tendency to Personalize

One important such factor is the impersonality of mathematics. Some people personalize events excessively, resisting an external perspective, and since numbers and an impersonal view of the world are intimately related, this resistance contributes to an almost willful innumeracy.

Quasi-mathematical questions arise naturally when one transcends one's self, family, and friends. How many? How long ago? How far away? How fast? What links this to that? Which is more likely? How do you integrate your projects with local, national, and international events? with historical, biological, geological, and astronomical time scales?

People too firmly rooted to the center of their lives find such questions uncongenial at best, quite distasteful at worst. Numbers and "science" have appeal for these people only if they're tied to them personally. They're often attracted to New Age beliefs such as Tarot cards, the *I Ching*, astrology and biorhythms, since these provide them with personally customized pronouncements. Getting such people interested in a numerical or scientific fact for its own sake or because it's intriguing or beautiful is almost impossible.

Though innumeracy may seem far removed from these people's real problems and concerns—money, sex, family, friends—it affects them (and all of us) directly and in many ways. If you walk down the main street of a resort town any summer night, for example, and see happy people holding hands, eating ice-cream cones, laughing, etc., it's easy to begin to think that other people are happier, more loving, more productive than you are, and so become unnecessarily despondent.

Yet it is precisely on such occasions that people display their good attributes, whereas they tend to hide and become "invisible" when they are depressed. We should all remember that our impressions of others are usually filtered in this way, and that our sampling of people and their moods is not random. It's beneficial to wonder occasionally what percentage of people you encounter suffer from this or that disease or inadequacy.

It's natural sometimes to confuse a group of individuals with some ideal composite individual. So many talents, so many different attractions, so much money, elegance, and beauty on display, but—and it's a trivial observation—this multitude of desiderata is inevitably spread out among a large group of people. Any given individual, no matter how brilliant or rich or attractive he or she is, is going to have serious shortcomings. Excessive concern with oneself makes it difficult to see this and thus can lead to depression as well as innumeracy.

Too many people, in my opinion, maintain a "Why me?" attitude toward their misfortunes. You needn't be a mathematician to realize that there's something wrong statistically if most people do this. It's like the innumerate high school principal who complains that most of his students score below his school's median SAT score. Bad things happen periodically, and they're going to happen to somebody. Why not you?

The Ubiquity of Filtering and Coincidence

Broadly understood, the study of filtering is nothing less than the study of psychology. Which impressions are filtered out and which are permitted to take hold largely determines our personality. More narrowly construed as the phenomenon whereby vivid and personalized events are remembered and their incidence therefore overestimated, the so-called Jeane Dixon effect often seems to lend support to bogus medical, diet, gambling, psychic, and pseudoscientific claims. Unless one is almost viscerally aware of this psychological tendency toward innumeracy, it is liable to bias our judgments.

As we've noted, a defense against this tendency is to look at bald numbers, to provide some perspective. Remember that rarity in itself leads to publicity, making rare events appear commonplace. Terrorist kidnappings and cyanide poisonings are given monumental coverage, with profiles of the distraught families, etc., yet the number of deaths due to smoking is roughly the equivalent of three fully loaded jumbo jets crashing each and every day of the year, more than 300,000 Americans annually. AIDS, as tragic as it is, pales in worldwide comparison to the more prosaic malaria, among other diseases. Alcohol abuse, which in this country is the direct cause of 80,000 to 100,000 deaths per year and a contributing factor in an additional 100,000 deaths, is by a variety of measures considerably more costly than drug abuse. It's not hard to think of other examples (famines and even genocides scandalously underreported), but it's necessary to remind ourselves of them periodically to keep our heads above the snow of media avalanches.

If one filters out banal and impersonal events, most of what's left are astounding aberrations and coincidences, and one's mind begins to resemble the headlines of supermarket tabloids.

Even people who have less restrictive filters and a good feel for numbers will note an increasingly large number of coincidences, due in large measure to the number and complexity of man-made conventions. Primitive man, in noticing the relatively few natural coincidences in his environment, slowly developed the raw observational data out of which science evolved. The natural world, however, does not offer immediate evidence for many such coincidences on its surface (no calendars, maps, directories, or even names). But in recent years the plethora of names and dates and addresses and organizations in a complicated world appears to have triggered many people's inborn tendency to note coincidence and improbability, leading them to postulate connections and forces where there are none, where there is only coincidence.

Our innate desire for meaning and pattern can lead us astray if we don't remind ourselves of the ubiquity of coincidence, an ubiquity which is the consequence of our tendency to filter out the banal and impersonal, of our increasingly convoluted world, and, as some of the earlier examples showed, of the unexpected frequency of many kinds of coincidence. Belief in the necessary or even probable significance of coincidences is a psychological remnant of our simpler past. It constitutes a kind of psychological illusion to which innumerate people are particularly prone.

The tendency to attribute meaning to phenomena governed only by chance is ubiquitous. A good example is provided by regression to the mean, the tendency for an extreme value of a random quantity whose values cluster around an average to be followed by a value closer to the average or mean. Very intelligent people can be expected to have intelligent offspring, but in general the offspring will not be as intelligent as the parents. A similar tendency toward the average or mean holds for the children of very short parents, who are likely to be short, but not as short as their parents. If I throw twenty darts at a target and manage to hit the bull's-eye eighteen times, the next time I throw twenty darts, I probably won't do as well.

This phenomenon leads to nonsense when people attribute the regression to the mean as due to some particular scientific law, rather than to the natural behavior of any random quantity. If a beginning pilot makes a very good landing, it's likely that his next one will not be as impressive. Likewise, if his landing is very bumpy, then, by chance alone, his next one will likely be better. Psychologists Amos Tversky and Daniel Kahneman studied one such situation in which, after good landings, pilots were praised, whereas, after bumpy landings, they were berated. The flight instructors mistakenly attributed the pilots' deterioration to their praise of them, and likewise the pilots' improvement to their criticism; both, however, were simply regressions to the more likely mean performance. Because this dynamic is quite general, Tversky and Kahneman write, "behavior is most likely to improve after punishment and to deteriorate after reward. Consequently, the human condition is such that . . . one is most often rewarded for punishing others, and most often punished for rewarding them." It's not necessarily the human condition, I would hope, but a remediable innumeracy which results in this unfortunate tendency.

The sequel to a great movie is usually not as good as the original. The reason may not be the greed of the movie industry in cashing in on the first film's popularity, but simply another instance of regression to the mean. A great season by a baseball player in his prime will likely be followed by a less impressive season. The same can be said of the novel after the best-seller, the album that follows the gold record, or the proverbial sophomore jinx. Regression to the mean is a widespread phenomenon, with instances just about everywhere you look. As mentioned in Chapter 2, however, it should be carefully distinguished from the gambler's fallacy, to which it bears a superficial resemblance.

Though chance fluctuations play a very large role in the price of a stock or even of the market in general, especially in the short term, the price of a stock is not a completely random walk, with a constant probability (P) of going up and a complementary probability (1-P) of going down, independent of its past performance. There is some truth to so-called fundamental analysis, which looks to the economic factors underlying a stock's value. Given that there is some rough economic estimate of a stock's value, regression to the mean can sometimes be used to justify a kind of contrarian strategy. Buy those stocks whose performance has been relatively lackluster for the previous couple of years, since they're more likely to regress to their mean and increase in price than are stocks which have performed better than their economic fundamentals would suggest and are therefore likely to regress to their mean and decline in price. A number of studies support this schematic strategy.

Decisions and Framing Questions

Judy is thirty-three, unmarried, and quite assertive. A magna cum laude graduate, she majored in political science in college and was deeply involved in campus social affairs, especially in anti-discrimination and anti-nuclear issues. Which statement is more probable?

(a) Judy works as a bank teller.

(b) Judy works as a bank teller and is active in the feminist movement.

The answer, surprising to some, is that (a) is more probable than (b), since a single statement is always more probable than a conjunction of two statements. That I will get heads when I flip this coin is more probable than that I will get heads when I flip this coin and get a 6 when I roll that die. If we have no direct evidence or theoretical support for a story, we find that detail and vividness vary inversely with likelihood; the more vivid details there are to a story, the less likely the story is to be true.

Getting back to Judy and her job at the bank, psychologically what may happen is that the preamble causes people to confuse the conjunction of statements of alternative (*b*) ("She's a teller and she's a feminist") with the conditional statement ("Given that she's a teller, she's probably also a feminist"), and this latter statement seems more probable than alternative (*a*). But this, of course, is not what (*b*) says.

Psychologists Tversky and Kahneman attribute the appeal of answer (*b*) to the way people come to probability judgments in mundane situations. Rather than trying to decompose an event into all its possible outcomes and then counting up the ones that share the characteristic in question, they form representative mental models of the situation, in this case of someone like Judy, and come to their conclusion by comparison with these models. Thus, it seems to many people that answer (*b*) is more representative of someone with Judy's background than is answer (*a*).

Many of the counter-intuitive results cited in this book are psychological tricks similar to the above, which can induce temporary innumeracy in even the most numerate. In their fascinating book *Judgement under Uncertainty*, Tversky and Kahneman describe a different variety of the seemingly irrational innumeracy that characterizes many of our most critical decisions. They ask people the following question: Imagine you are a general surrounded by an overwhelming enemy force which will wipe out your 600-man army unless you take one of two available escape routes. Your intelligence officers explain that if you take the first route you will save 200 soldiers, whereas if you take the second route the probability is ⅓ that all 600 will make it, and ⅔ that none will. Which route do you take?

Most people (three out of four) choose the first route, since 200 lives can definitely be saved that way, whereas the probability is ⅔ that the second route will result in even more deaths.

So far, so good. But what about the following? Again, you're a general faced with a decision between two escape routes. If you take the first one, you're told, 400 of your soldiers will die. If you choose the second route, the probability is ⅓ that none of your soldiers will die, and ⅔ that all 600 will die. Which route do you take?

Most people (four out of five) faced with this choice opt for the second route, reasoning that the first route will lead to 400 deaths, while there's at least a probability of ⅓ that everyone will get out okay if they go for the second route.

The two questions are identical, of course, and the differing responses are a function of how the question is framed, whether in terms of lives saved or of lives lost.

Another example from Tversky and Kahneman: Choose between a sure $30,000 or an 80 percent chance of winning $40,000 and a 20 percent chance of winning nothing. Most people will take the $30,000 even though the average expected gain in the latter choice is $32,000 (40,000 × .8). What if the choices are either a sure loss of $30,000 or an 80 percent chance of losing $40,000 and a 20 percent chance of losing nothing? Here most people will take the chance of losing $40,000 in order to have a chance of avoiding any loss, even though the average expected loss in the latter choice is $32,000 (40,000 × .8). Tversky and Kahneman conclude that people tend to avoid risk when seeking gains, but choose risk to avoid losses.

Of course, we needn't resort to such clever examples to realize that how a question or statement is framed plays a big role in how someone responds to it. If you asked an average taxpayer how he feels about a 6 percent utility increase, he'd probably be amenable. If you asked his reaction to a $91 million hike in utility bills, he probably wouldn't be. Saying that someone scored in the middle third of his class is more impressive than saying that he scored in the 37th percentile (better than 37 percent of his peers).

Math Anxiety

A more common source of innumeracy than psychological illusions is what Sheila Tobias calls math anxiety. In *Overcoming*

Math Anxiety she describes the block many people (especially women) have to any kind of mathematics, even arithmetic. The same people who can understand the subtlest emotional nuances in conversation, the most convoluted plots in literature, and the most intricate aspects of a legal case can't seem to grasp the most basic elements of a mathematical demonstration.

They seem to have no mathematical frame of reference and no basic understandings on which to build. They're afraid. They've been intimidated by officious and sometimes sexist teachers and others who may themselves suffer from math anxiety. The infamous word problems terrify them, and they're convinced that they're dumb. They feel that there are mathematical minds and nonmathematical minds, and that the former always come up with answers instantaneously whereas the latter are helpless and hopeless.

Not surprisingly, these feelings constitute a formidable block to numeracy. There are things to be done for those who suffer from them, however. One simple technique which works surprisingly well is to explain the problem clearly to someone else; if a person can sit still for this, he or she may think about the problem long enough to realize that additional thought might bring results. Other techniques may be: to use smaller numbers; to examine related but easier problems or sometimes related but more general problems; to collect information relevant to the problem; to work backwards from the solution; to draw pictures and diagrams; to compare the problem or parts of it to problems you do understand; and most important of all, to study as many different problems and examples as possible. The truism that one learns how to read by reading and how to write by writing extends to solving mathematical problems (and even to constructing mathematical proofs).

In writing this book, I've come to understand a way in which I (and probably mathematicians in general) contribute unintentionally to innumeracy. I have a difficult time writing at extended length about anything. Either my mathematical training or my natural temperament causes me to distill the crucial points and not to dwell (I want to write "dither") over side issues or contexts or biographical detail. The result, I think, is clean exposition, which can nevertheless be intimidating to people who expect a more leisurely approach. The solution is for a variety of people to write about mathematics. As has been said about many subjects, mathematics is too important to be left to the mathematicians.

Different from and much harder to deal with than math anxiety is the extreme intellectual lethargy which affects a small but growing number of students, who seem to be so lacking in mental discipline or motivation that nothing can get through to them. Obsessive-compulsive sorts can be loosened up and people suffering from math anxiety can be taught ways to allay their fears, but what about students who don't care enough to focus any of their energy on intellectual matters? You remonstrate: "The answer's not X but Y. You forgot to take account of this or that." And the response is a blank stare or a flat "Oh, yeah." Their problems are an order of magnitude more serious than math anxiety.

Romantic Misconceptions

Romantic misconceptions about the nature of mathematics lead to an intellectual environment hospitable to and even encouraging of poor mathematical education and psychological distaste for the subject and lie at the base of much innumeracy. Rousseau's disparagement of the English as "a nation of shopkeepers" persists as a belief that a concern with numbers and details numbs one to the big questions, to the grandeur of nature. Mathematics is often taken to be mechanical, the work of low-level technicians who will report to the rest of us anything we absolutely must know. Alternatively, mathematics is sometimes endowed with a coercive character which is somehow capable of determining our future. Attitudes such as these certainly predispose one to innumeracy. Let's examine some of them.

Mathematics is thought to be cold, since it deals with abstraction and not with flesh and blood. In a sense, of course, this is true. Even Bertrand Russell termed the beauty of pure mathematics "cold and austere," and it is precisely this cold and austere beauty that initially attracts mathematicians to the subject, since most are essentially Platonists and conceive of mathematical objects as existing in some abstract, ideal realm.

Still, pure mathematics is only part of the story; almost equally important is the interplay between these ideal Platonic forms (or whatever they are) and their possible interpretation in the real world, and in this extended sense mathematics is not cold. Recall that mathematics as simple as "1 + 1 = 2" can still be thoughtlessly misapplied: If 1 cup of popcorn is added to 1 cup of water, we do not end up with 2 cups of soggy popcorn. In trivial cases as well as in difficult ones, mathematical applications can be a tricky affair, requiring as much human warmth and nuance as any other endeavor.

Even when mathematics is at its purest and coldest, its pursuit is often quite hot. Like any other scientists, mathematicians are motivated by a complex of emotions including healthy doses of jealousy, arrogance, and competitiveness. Research mathematicians attack their problems with an intensity and compulsiveness which seem related to the purity of their research. A strong streak of romanticism runs through mathematics, manifesting itself most clearly in the most fundamental areas of mathematics, number theory and logic. This romanticism extends at least as far back as the mystical Pythagoras, who believed that the secret of understanding the world lay in understanding number; it found expression in the numerology and Cabala of the Middle Ages, and persists (in a nonsuperstitious form) in the Platonism of the modern logician Kurt Gödel and others. The existence of this romantic tendency constitutes at least a small part of the emotional makeup of most mathematicians, and is perhaps surprising to those who think of mathematicians as cold rationalists.

Another widespread misconception is that numbers depersonalize or somehow diminish individuality. There is of course a legitimate concern about reducing complicated phenomena to simple numerical scales or statistics. Fancy mathematical terms and reams of statistical correlations and computer printouts do not in themselves produce understanding, social scientists' claims notwithstanding. Reducing a complex intelligence or the economy to numbers on a scale, whether I.Q. or GNP, is myopic at best and many times simply ludicrous.

That said, objections to being identified for special purposes by number (social security, credit cards, etc.) seem silly. If anything, a number in these contexts enhances individuality; no two people have the same credit-card number, for example, whereas many have similar names or personality traits or socioeconomic profiles. (I personally use my middle name—John Allen Paulos—to keep the masses from confusing me with the Pope.)

I'm always amused by commercials for banks which tout their personalized service, which service amounts to a poorly trained and badly paid cashier saying "Good morning" and then promptly fouling up your transaction. I'd rather go to a machine which knows me by some code word but on whose operating programs a team of software writers has painstakingly worked for months.

One objection I do have to identification numbers is their excessive length. An application of the multiplication principle shows that a nine-digit number or a six-letter sequence is of more than adequate length to distinguish every person in the country (10^9 is a billion, while 26^6 is more than 300 million). Why do department stores or suburban water companies find it necessary to assign account numbers with twenty or more symbols?

Writing of numbers and individuation reminds me of companies which will name a star after anyone who pays a $35 fee. So that the companies can wrap themselves in some sort of official cloak, the names are recorded in books which are registered in the Library of Congress. The companies generally advertise around Valentine's Day, and judging from their longevity they must be doing a fairly good business. A related and equally silly business idea I had was to "officially" associate a number with anyone who pays a $35 fee. A certificate would be sent to subscribers and a book with their names and cosmic numbers would be registered in the Library of Congress. There might even be a sliding scale, with perfect numbers selling at a premium and prime numbers going for more than nonperfect composite numbers, etc. I would get rich selling numbers.

Yet another misperception people have of mathematics is that it is limiting, and somehow opposed to human freedom. If they accept certain statements and then are shown that some other unpleasant statements follow from them, they associate the unpleasantness of the conclusions with the vehicle of their expression.

In this very weak sense, of course, mathematics is constraining, as is all reality, but it has no independent power to coerce. If one accepts the premises and definitions, one must accept what follows from them, but one can frequently reject premises or refine definitions or choose a different mathematical approach. In this sense, mathematics is just the opposite of constraining; it is empowering, and at the service of anyone who cares to use it.

Consider the following example, which illustrates the way we use mathematics but are not bound by it. Two men bet on a series of coin flips. They agree that the first to win six such flips will be awarded $100. The game, however, is interrupted after only eight flips, with the first man leading 5 to 3. The question is: How should the pot be divided? One might say that the first man should be awarded the full $100, because the bet was all or nothing and he was leading. Or one could reason instead that the first man should receive $\frac{5}{8}$ of the pot and the other one the remaining $\frac{3}{8}$ because the score was 5 to 3. Alternatively, it might be argued that because the probability of the first man's going on to win can be computed to be $\frac{7}{8}$ (the only way the second man could have won is by winning three flips in a row, a feat with probability $\frac{1}{8} = \frac{1}{2} \times \frac{1}{2} \times \frac{1}{2}$), the first man should receive $\frac{7}{8}$ of the pot to $\frac{1}{8}$ for the second man. (This incidentally was Pascal's solution to this, one of the first problems in probability theory.) Rationales for still other ways to apportion the money are possible.

The point is that the criteria for deciding on any one of these divisions are nonmathematical. Mathematics can help determine the consequences of our assumptions and values, but we, not some mathematical divinity, are the origin of these assumptions and values.

Nevertheless, mathematics is often seen as a spiritless affair. Many people believe that determining the truth of any mathematical statement is merely a matter of mechanically plugging into some algorithm or recipe which will eventually yield a yes or a no answer, and that given a reasonable collection of basic axioms, every mathematical statement is either provable or unprovable. Mathematics in this view is cut-and-dried and calls for nothing so much as mastery of the requisite algorithms, and unlimited patience.

The Austrian–American logician Kurt Gödel brilliantly refuted these facile assumptions by demonstrating that any system of mathematics, no matter how elaborate, will of necessity contain statements which can be neither proved nor disproved within the system. This and related results by logicians Alonzo Church, Alan Turing, and others have deepened our understanding of mathematics and its limitations. Given our concerns here, however, it's sufficient to note that not even theoretically is mathematics mechanical or complete.

Though related to these abstract considerations, the mistaken belief in the mechanical nature of mathematics generally takes more prosaic forms. Mathematics is often viewed as a subject for technicians, and mathematical talent is confused with rote skills, elementary programming ability, or speed of calculation. In a curious way, many people simultaneously exalt and dismiss mathematicians and scientists as impractical whizzes. Consequently, we frequently find senior mathematical, engineering, and scientific people ardently wooed by industry and then subordinated to newly minted M.B.A.s and accountants.

One other prejudice people have about mathematics is that its study somehow diminishes one's feeling for nature and the "big" questions. Since this position is frequently expressed (for example, by Whitman at the beginning of this chapter), but rarely argued for, it's difficult to refute. It makes as much sense as believing that a technical knowledge of molecular biology will render a person unappreciative of the mysteries and complexities of life. Too often, this concern for the big picture is simply obscurantist and is put forward by people who prefer vagueness and mystery to (partial) answers. Vagueness is at times necessary and mystery is never in short supply, but I don't think they're anything to worship. Genuine science and mathematical precision are more intriguing than are the "facts" published in supermarket tabloids or a romantic innumeracy which fosters credulity, stunts skepticism, and dulls one to real imponderables.

Digression: A Logarithmic Safety Index

Several years ago, supermarkets began to use unit pricing (cents per pound, per fluid ounce, etc.) to give consumers a uniform scale with which to measure value. If the price of dog food and cake mix can be rationalized, why can't some sort of rough "safety index" be devised which allows us to gauge how safe various activities, procedures, and illnesses are? What I'm suggesting is a kind of Richter scale which the media could use as a shorthand for indicating degrees of risk.

Like the Richter scale, the proposed index would be logarithmic, so what follows is a little detour to review for the innumerate those dreaded monsters from high school algebra: logarithms. The logarithm of a number is simply the power to which 10 must be raised to equal the number in question. The logarithm of 100 is 2 because $10^2 = 100$; the logarithm of 1,000 is 3 because $10^3 = 1,000$; and the logarithm of 10,000 is 4 because $10^4 = 10,000$. For numbers between powers of 10, the logarithm is between the two nearest powers of 10. For example, the logarithm of 700 is between 2, the logarithm of 100, and 3, the logarithm of 1,000; it happens to be about 2.8.

The safety index would work as follows. Consider some activity which results in a certain number of deaths per year; say, automobile driving: one American in 5,300 dies each year due to a car crash. The safety index associated with automobile driving is thus a relatively low 3.7, the logarithm of 5,300. More generally, if one person out of X dies as a result of some given activity each year, the safety index for that activity is simply the logarithm of X. Thus, the higher the safety index, the safer the activity in question.

(Since people and the media are sometimes more interested in danger than in safety, an alternative approach might be to define a danger index equal to 10 minus the safety index. A 10 on such a danger index would then correspond to a safety index of 0—certain death; and a low danger index of 3 would be equivalent to a high safety index of 7, or one chance in 10^7 of dying.)

According to the Centers for Disease Control, smoking results in an estimated 300,000 premature deaths annually in the United States, which is equivalent to one American in 800 dying each year due to heart, lung, and other diseases caused by smoking. The logarithm of 800 is 2.9, and thus the safety index for smoking is even lower than that for driving. A more graphic way to describe the number of such preventable deaths is to note that each year seven times as many people die from cigarette smoking as were killed in the entire war in Vietnam.

Automobile driving and smoking have safety indices of 3.7 and 2.9, respectively. Contrast these small values with the safety index of being kidnapped. It's estimated that fewer than 50 American children are kidnapped each year by strangers, and the incidence of kidnapping is thus about one in 5 million, resulting in a safety index of 6.7. Remember that the bigger the number, the smaller the risk, and that for every unit increase in the safety index, the risk declines by a factor of 10.

The virtue of such a coarse logarithmic safety scale is that it provides us, and particularly the media, with an order-of-magnitude estimate of the risks associated with various activities, illnesses, and procedures. There is a possible problem, however, since the index doesn't clearly distinguish incidence from likelihood. An activity may be very dangerous but quite rare, and thus would result in few deaths and have a high safety index. For example, few people die as a result of high-wire acrobatics between skyscrapers, an activity which is nevertheless not safe.

A slight refinement must therefore be made in the index, by considering only those people who are likely to engage in the activity in question. If one out of every X of them dies due to the activity, then the safety index of the activity would be the logarithm of X. On this basis, the safety index of high-wire acrobatics between skyscrapers might be a very low 2 (estimating that one out of every 100 such daredevil acrobats doesn't make it across). Likewise, playing Russian roulette each year (one chamber in six containing a bullet) has a safety index of less than 1, approximately 0.8.

Activities or illnesses whose safety indices are greater than 6 should be considered quite safe, being equivalent to less than one chance in a million per year. Anything with a safety index of less than 4 should be considered warily, being equivalent to a more than one chance per 10,000 per year. Publicity, of course, tends to obscure these numbers, but like the Surgeon General's warning on cigarette packages, the figures would eventually begin to filter into the public consciousness. Victim-oriented reporting would have a less misleading impact if the safety index were kept firmly in mind. Isolated but vivid tragedies involving a few people should not blind us to the fact that myriad prosaic activities may involve a much higher degree of risk.

Let's look at a few more examples. The 12,000 American deaths due to heart and circulatory disease each week translate into an annual rate of one death per 380, for a safety index of 2.6. (If one is a nonsmoker, the safety index for heart and circulatory diseases is considerably higher, but we're only interested in shorthand approximations here.) The safety index for cancer is a slightly better 2.7. An activity in the marginal area is bike riding: one American in 96,000 dies each year in a bicycle crash, for a safety index of about 5 (actually, somewhat lower, since relatively few people ride bikes). In the rare category, it's estimated that one American in 2,000,000 is killed by lightning, for a safety index of 6.3; whereas one in 6,000,000 dies from a bee sting each year, for a safety index of 6.8.

The safety index varies over time, death from influenza and pneumonia going from a safety index of approximately 2.7 in 1900 to approximately 3.7 in 1980. Over the same period, the risk of death from tuberculosis went from 2.7 to approximately 5.8. Variations between countries are to be expected—the safety index for homicide being approximately 4 in the United States and between 6 and 7 in Great Britain, whereas malaria's index is orders of magnitude lower in most of the world than it is in the United States. Comparable economies of expression can be obtained by comparing the high safety index associated with nuclear power with the relatively low safety index for burning coal.

In addition to the ready perspective it offers on relative risk, the safety index underscores the obvious fact that every activity carries some risk. It provides the rough answer to the crucial question: How much?

Whatever the merits of such a safety index, I think the establishment of statistical ombudsmen by television networks, news magazines, and major newspapers would be a welcome and effective step in combating innumeracy in the media. An ombudsman would scan the news stories, research whatever statistics are mentioned, try to see that they are at least internally consistent, and probe most carefully into a priori implausible claims. Perhaps a regular feature similar to William Safire's *New York Times* column on usage might consider the worst innumeracies of the week or month. It would have to be quite entertainingly written, however, since, though there's happily a small army of readers interested in verbal niceties, relatively few are interested in similar but often more important numerical nuances.

These matters are not merely academic, and there is a direct way in which the mass media's predilection for dramatic reporting leads to extreme politics and even pseudoscience. Since fringe politicians and scientists are generally more intriguing than mainstream ones, they garner a disproportionate share of publicity and thus seem more representative and significant than they otherwise would. Furthermore, since perceptions tend to become realities, the natural tendency of the mass media to accentuate the anomalous, combined with an innumerate society's taste for such extremes, could conceivably have quite dire consequences.

5

Statistics, Trade-Offs,
and Society

There was once a state legislator in Wisconsin who objected to the introduction of daylight saving time despite all the good arguments for it. He maintained sagely that there is always a trade-off involved in the adoption of any policy, and that if daylight saving time were instituted, curtains and other fabrics would fade more quickly.

Sixty-seven percent of the doctors surveyed preferred X to Y. (Jones couldn't be persuaded.)

It's been estimated that, because of the exponential growth of the world's population, between 10 and 20 percent of all the human beings who have ever lived are alive now. If this is so, does this mean that there isn't enough statistical evidence to conclusively reject the hypothesis of immortality?

Priorities—Individual vs. Societal

This chapter will concentrate on the harmful social effects of innumeracy, with particular emphasis on the conflict between society and the individual. Most of the examples consider some form of trade-off or balancing of conflicting concerns, and will show how innumeracy contributes to the relative invisibility of these trade-offs, or sometimes, as in the case of the Wisconsin legislator above, to seeing them where they aren't.

Let's examine a preliminary and relevant probability oddity, whose discovery is due to statistician Bradley Efron. Imagine four dice, A, B, C, and D, strangely numbered as follows: A has 4 on four faces and 0 on two faces; B has 3s on all six faces; C has four faces with 2 and two faces with 6; and D has 5 on three faces and 1 on three faces.

If die A is rolled against die B, die A will win—by showing a higher number—two-thirds of the time; similarly, if die B is rolled against die C, B will win two-thirds of the time; if die C is rolled against die D, it will win two-thirds of the time; nevertheless, and here's the punch line, if die D is rolled against die A, it will win two-thirds of the time. A beats B beats C beats D beats A, all two-thirds of the time. You might even profit from this by challenging someone to choose whatever die he or she wanted, and you could then choose a die which would beat it two-thirds of the time. If they choose die B, you choose A; if they choose A, you choose D; and so on.

That die C beats die D may require some explanation. Half of the time, a 1 will turn up on die D, in which case die C will certainly win. The other half of the time, a 5 will turn up on die D, in which case die C will win one-third of the time. Thus, since C can win in these two different ways, it beats D exactly $\frac{1}{2} + (\frac{1}{2} \times \frac{1}{3}) = \frac{2}{3}$ of the time. A similar argument can be used to show that die D beats die A two-thirds of the time. This kind of nontransitivity (where X beats Y and Y beats Z and Z beats W, but W nevertheless beats X) is at the base of most voting paradoxes, from the Marquis de Condorcet in the eighteenth century to Kenneth Arrow in the twentieth.

The possibility of social irrationality resting on a base of individual rationality is suggested by a slight variant of Condorcet's original example. In it there are three candidates for public office, whom I'll call Dukakis, Gore, and Jackson, to commemorate the 1988 Democratic primary battles. Assume that one-third of the electorate prefers Dukakis to Gore to Jackson, that another one-third prefers Gore to Jackson to Dukakis, and that the last one-third prefers Jackson to Dukakis to Gore. So far, so good.

But if we examine the possible two-man races, a paradox appears. Dukakis will boast that two-thirds of the electorate prefer him to Gore, whereupon Jackson will respond that two-thirds of the electorate prefer him to Dukakis. Finally, Gore will counter by noting that two-thirds of the electorate prefer him to Jackson. If societal preferences are determined by majority vote, "society" prefers Dukakis over Gore, Gore over Jackson, and Jackson over Dukakis. Thus, even if the preferences of all the individual voters are rational (i.e., transitive—whenever a voter prefers X to Y and Y to Z, then that voter prefers X to Z), it doesn't necessarily follow that the societal preferences determined by majority rule are transitive, too.

Of course, in real life, things can get considerably more complex. Mort Sahl remarked about the 1980 election, for example, that people weren't so much voting for Reagan as they were voting against Carter, and that if Reagan had run unopposed he would have lost. (I don't know how to model that situation.)

One should not get the mistaken impression that Condorcet's paradox and Sahl's joke are equally unrealistic. The economist Kenneth Arrow has proved a very powerful generalization which shows that something like the above situation characterizes every voting system. Specifically, he demonstrated that there is never a way to derive societal preferences from individual preferences that can be absolutely guaranteed to satisfy these four minimal conditions: the societal preferences must be transitive; the preferences (individual and societal) must be restricted to available alternatives; if every individual prefers X to Y, then the societal preference must be for X over Y; and no individual's preferences automatically determine the societal preferences.

Laissez-Faire: Adam Smith or Thomas Hobbes

A different sort of conflict between the individual and society is revealed in a dilemma devised by the logician Robert Wolf that is related to the more famous prisoner's dilemma, to which we'll return shortly. Both demonstrate that acting in one's self-interest does not always best serve one's self-interest.

Imagine that you and twenty casual acquaintances are in a room together, having been brought there by an eccentric philanthropist. None of you can communicate in any way with one another, and you're each given the choice of either pressing a small button in front of you or not.

If all of you refrain from pressing the button, you'll each receive $10,000 from the philanthropist. But if at least one of you presses the button, those of the group who press the button will receive $3,000 and those who refrain from pressing the button will receive nothing. The question is, do you press the button for a sure $3,000, or refrain and hope that everybody else in the group does the same, so that you each get $10,000.

Whatever your decision, one can vary the stakes or the number of people involved so as to induce you to reverse your decision. If you decided to press the button, you probably would have reversed your decision if the stakes had been $100,000 vs. $3,000. If you refrained from pressing, you would probably have reversed that decision if the stakes had been $10,000 vs. $9,500.

There are other ways of raising the stakes. Replace the eccentric philanthropist with a powerful sadist. If no member of the group presses the button, he'll allow each of you to leave safely. However, if some of you do press the button, the ones who do will be forced by the sadist to play Russian roulette, with a 95 percent chance of survival, while the ones who don't will be killed outright. Do you press the button and take the 95 percent chance of survival, and assume the cost of indirectly leading to the deaths of others, or do you resist your fear and not press the button and hope that no one else's fear gets the better of him?

Wolf's dilemma often arises in situations where we fear we're going to be left behind if we don't watch out for ourselves.

Now consider the case of two women who must make a brief, hurried transaction (let's suppose they're drug traffickers). The women exchange filled brown-paper bags on a street corner and depart quickly before checking the contents of the bag each has received. Before the meeting, each has the same option: to put in her bag the item of worth which the other wants (the cooperative option) or to fill it with shredded newspaper (the individualist option). If they cooperate with each other, each will receive what she wants, but at a fair cost. If A fills her bag with shredded newspaper and B doesn't, A will get what she wants at no cost and B will be duped. If they both fill their bags with shredded newspaper, neither will get what she wants, but neither will be duped.

The best outcome for the women as a pair is for them to cooperate with each other. A, however, can reason as follows: If B takes the cooperative option, I can get what I want at no cost to me by taking the individualist option. On the other hand, if B takes the individualist option, at least I won't be duped if I do, too. Thus, regardless of what B does, I'm better off if I take the individualist alternative and give her a bag full of newspaper. B can, of course, reason in the same way, and they're both likely to end up exchanging worthless bags of shredded newspaper.

A similar situation can arise in legitimate business transactions or, indeed, in almost any sort of exchange.

The prisoner's dilemma owes its name to a scenario, formally identical to the one above, wherein two men suspected of a major crime are apprehended in the course of committing some minor offense. They're separated and interrogated, and each is given the choice of confessing to the major crime and implicating his partner or remaining silent. If they both remain silent, they'll each get only one year in prison. If one confesses and the other doesn't, the one who confesses will be rewarded by being let go, while the other one will get a five-year term. If they both confess, they can both expect to spend three years in prison. The cooperative option is to remain silent, while the individualist option is to confess.

The dilemma, again, is that what's best for them as a pair, to remain silent and spend a year in prison, leaves each of them open to the worst possibility, being a patsy and spending five years in prison. As a result, they'll probably both confess and both spend three years in prison.

So what? The appeal of the dilemma has nothing to do, of course, with any interest we might have in women drug traffickers or in the criminal justice system. Rather, it provides the logical skeleton for many situations we face in everyday life. Whether we're businessmen in a competitive market or spouses in a marriage or superpowers in an arms race, our choices can often be phrased in terms of the prisoner's dilemma. There isn't always a right answer, but the parties involved will be better off as a pair if each resists the temptation to double-cross the other and instead cooperates with or remains loyal to him or her. If both parties pursue their own interests exclusively, the outcome is worse than if both cooperate. Adam Smith's invisible hand ensuring that individual pursuits bring about group well-being is in these situations quite paralyzed.

A somewhat different situation is that of two authors who must publicly review each other's book. If they appeal to the same limited audience, there is a certain payoff to panning the other's book while one's own book is praised, and this individual payoff is greater than that resulting from mutual praise, which in turn is greater than a mutual panning. Thus, the choice of whether to praise or to pan is something of a prisoner's dilemma. (I say "something of" because there should be more weighty considerations, such as the merit of the books in question.)

There is an extensive literature on the subject of prisoner's dilemmas. The two-party prisoner's dilemma can be extended to a situation where there are many people, each having the choice whether to make a minuscule contribution to the public good or a massive one to his own private gain. This many-party prisoner's dilemma is useful in modeling situations where the economic value of "intangibles" such as clean water, air, and space is an issue.

In another variation, political scientist Robert Axelrod has studied the iterated prisoner's dilemma situation wherein our two women drug traffickers (or businessmen or spouses or superpowers or whatever) meet again and again to make their transaction. Here there is a very compelling reason to cooperate with and not try to double-cross the other party: you're probably going to have to do business with him or her again.

Since, to a considerable extent, almost all social transactions have an element of the prisoner's dilemma in them, the character of a society is reflected in which such transactions lead to cooperation between parties and which don't. If the members of a particular "society" never behave cooperatively, their lives are likely to be, in Thomas Hobbes's words, "solitary, poor, nasty, brutish and short."

Birthdays, Death Days, and ESP

Probability theory began with gambling problems in the seventeenth century, and something of the gaming flavor and appeal clings to it to this day. Statistics began in the same century with the compilation of mortuary tables, and something of its origins sticks to it as well. Descriptive statistics, the oldest part of the subject and the part with which people are most familiar, is at times (though not always) a dreary discipline, with its ceaseless droning about percentiles, averages, and standard deviations. The theoretically more interesting field of inferential statistics uses probability theory to make predictions, to estimate important characteristics of a population, and to test the validity of hypotheses.

The latter notion—statistical testing of hypotheses—is simple in principle. You make an assumption (often, forbiddingly termed the null hypothesis), design and perform an experiment, then calculate to see if the results of the experiment are sufficiently probable, given the assumption. If they aren't, you throw out the assumption, sometimes provisionally accepting an alternative hypothesis. In this sense, statistics is to probability as engineering is to physics—an applied science based on a more intellectually stimulating foundational discipline.

Consider this example, in which the unexpected outcome of a simple statistical test is warrant enough to reject a common and seemingly obvious hypothesis: that people's birthdays and death days have no relation to each other. Specifically, it's natural to assume that approximately 25 percent of the deaths in a given community will occur within a three-month period following the decedents' birthdays (and 75 percent during the other nine months).

Surprisingly, however, a random sample of 747 death notices published in Salt Lake City, Utah, newspapers during 1977 indicated that 46 percent of the decedents surveyed died within the three-month period following their birthdays. Given the null hypothesis in question, that approximately 25 percent of the decedents would have died in the three-month interval after their birthdays, the probability that 46 percent or more would die during these intervals can be computed to be so tiny as to be practically zero. (We must consider the alternative hypothesis to be that 46 percent or more would die, and not that exactly 46 percent would die. Why?)

Thus, we can reject the null hypothesis and tentatively accept that, for whatever reason, people do seem to wait until their birthdays to die. Whether this is due to the desire to achieve another milestone or to the trauma of the birthday ("Oh, my God, I'm ninety-two!"), it seems clear that a person's psychological state is a factor affecting when he will die. It would be interesting to see this study replicated in a different city. My guess is that the phenomenon is more pronounced among very old people, for whom a last birthday may be the only kind of significant achievement within reach.

To illustrate the important binomial probability model, and to provide a numerical example of a statistical test, imagine the following miniature test for ESP. (This is one of the passages I mentioned that may be easily ignored.) Assume that one of three possible symbols chosen at random is placed under a piece of cardboard and the subject is asked to identify it. Over the course of twenty-five such trials, the subject correctly identifies the symbol ten times. Is this enough evidence to warrant rejecting an assumption that the subject does not have ESP?

The answer lies in determining the probability of doing this well or better by chance. The probability of making exactly ten correct guesses by chance is $(\frac{1}{3})^{10}$ (the probability of answering the first ten questions correctly) $\times$ $(\frac{2}{3})^{15}$ (the probability of answering the next fifteen questions incorrectly) $\times$ the number of different ten-question collections of the twenty-five-question test there are. This latter factor is needed, since we're interested in the probability that ten questions are answered correctly, not necessarily the first ten. Any collection of ten correct responses and fifteen incorrect responses is acceptable and has the same probability, $(\frac{1}{3})^{10} \times (\frac{2}{3})^{15}$.

Since the number of ways of choosing ten out of twenty-five is 3,628,800 [$(25 \times 24 \times 23 \ldots 17 \times 16)/(10 \times 9 \times 8 \times \ldots 2 \times 1)$], the probability of guessing correctly some ten out of twenty-five is $3,628,800 \times (\frac{1}{3})^{10} \times (\frac{2}{3})^{15}$. Similar calculations can be performed for eleven, twelve, thirteen, up to twenty-five correct responses out of twenty-five, and if these probabilities are added up, we get the probability of guessing at least ten out of twenty-five by chance—about 30 percent. This probability is not even close to being sufficiently low to warrant rejecting our assumption of no ESP. (Sometimes the outcomes are more difficult to dismiss probabilistically, but in these cases there have always been flaws in the experimental design which have provided the subject with cues.)

Type I and Type II Errors: From Politics to Pascal's Wager

One more example of a statistical test. Suppose I hypothesize that at least 15 percent of the cars in a certain region are

Corvettes, and upon watching one thousand cars go by representative intersections in the region note only eighty Corvettes among them. Using probability theory, I calculate that, given my assumption, the likelihood of this result is well below 5 percent, a commonly used "level of significance." Therefore I reject my hypothesis that 15 percent of the cars in the region are Corvettes.

There are two sorts of errors that can be made in applying this or any statistical test; they're called, imaginatively enough, Type I and Type II errors. A Type I error occurs when a true hypothesis is rejected, and a Type II error occurs when a false hypothesis is accepted. Thus, if a large number of Corvettes from a car show drove through the region and we therefore accepted the false assumption that at least 15 percent of the cars in the region were Corvettes, we would be making a Type II error. On the other hand, if we didn't realize that most of the Corvettes in the region weren't driven but were kept in garages, then in rejecting the true assumption we would be making a Type I error.

The distinction can also be applied less formally. When money is being distributed, the stereotypical liberal tries especially hard to avoid Type I errors (the deserving not receiving their share), whereas the stereotypical conservative is more concerned with avoiding Type II errors (the undeserving receiving more than their share). When punishment is being meted out, the stereotypical conservative is more concerned with avoiding Type I errors (the deserving or guilty not receiving their due), whereas the stereotypical liberal worries more about avoiding Type II errors (the undeserving or innocent receiving undue punishment).

Of course, there are always people who will object to the strictness of the Federal Drug Administration in not releasing drug X soon enough to prevent suffering, and also complain loudly when drug Y is released prematurely and causes severe complications. Like the FDA, which must evaluate the relative probabilities of a Type II error (okaying a bad drug) and a Type

I error (not okaying a good drug), we must constantly evaluate analogous probabilities for ourselves. Should we sell the rising stock option and risk losing out on its further ascent, or hold on to it and risk its decline and the loss of our premium? Should we operate, or manage medically? Should Henry ask Myrtle out and risk her saying no, or should he not and keep his peace of mind but not learn that she would have said yes?

Similar considerations apply to the manufacturing process. Often, after some crucial bit of machinery breaks down because of bad parts, or after some unusually unreliable string of items (firecrackers, cans of soup, computer chips, condoms) comes to light, there are calls for new controls to ensure that no more defectives are manufactured. This sounds reasonable, but in most cases it's simply impossible or, what amounts to the same thing, prohibitively expensive. There are quality-control checks whereby samples of each batch of manufactured goods are tested to ensure that there are no or very few defectives in the sample, but not every item is tested (or even testable).

There's almost always a trade-off between quality and price, between Type II errors (accepting a sample with too many defectives) and Type I errors (rejecting a sample with very few defectives). Moreover, if this trade-off is not acknowledged, there is a tendency to deny or cover up the inevitable defective items, which makes the job of quality control that much more difficult. Apropos of this is the proposed Strategic Defense Initiative, whose computer software, satellites, mirrors, etc., would be so awesomely complex that it takes a kind of innumerate naïveté to believe it will work without bankrupting the treasury.

The Strategic Defense Initiative brings thoughts of destruction and salvation, and even here the notion of trade-offs can play a useful role. Pascal's wager on God's existence, for example, can be cast as a choice between the relative probabilities and consequences of Type I and Type II errors. Should we accept God and act accordingly and risk a Type II error (He doesn't exist), or should we reject God and act accordingly and risk a Type I error (He does exist)? Of course, there are as-

sumptions underlying these phrases which are invalid or meaningless without clarification. Still, the point is that all kinds of
decisions can be cast into this framework and call for the informal
evaluation of probabilities. There is no such thing as a free lunch,
and even if there were, there'd be no guarantee against in-
digestion.

Polling with Confidence

Estimating characteristics of a population, such as the percentage who favor a certain candidate or a particular brand of
dog food, is, like hypothesis testing, simple in principle. One
selects a random sample (easier said than done) and then determines what percentage of the sample favors the candidate
(say, 45 percent) or the brand of dog food (say, 28 percent),
which percentages are then taken to be estimates of the opinion
of the population as a whole.

The only actual poll I ever took myself was informal and
designed to answer the burning question: What percentage of
college women enjoy watching the Three Stooges? Eliminating
those unfamiliar with the Stooges' slapstick, physical, lowbrow
comedy, I found that an overwhelming 8 percent of my sample
admitted to such an indulgence.

The care devoted to the selection of the above sample was
not great, but at least the result, 8 percent, had a credible ring
to it. One obvious problem with statements such as "67 percent
(or 75 percent) of those surveyed favored tablet X" is that they
may be based on tiny samples of three or four. Even more
extreme is the case where a celebrity endorses a diet or medicine
or whatever, in which case we have a sample of one, and generally a paid sample at that.

Thus, more difficult than making statistical estimates is deciding how much confidence we should place in them. If the
sample is large, we can have more confidence that its characteristics are close to those of the population as a whole. If the
distribution of the population is not too dispersed or varied, we

can, again, have more confidence that the sample's characteristics are representative.

By utilizing a few principles and theorems in probability and statistics, we can come up with so-called confidence intervals to estimate how likely a sample characteristic is to be representative of the population as a whole. Thus, we might say that a 95 percent confidence interval for the percentage of voters favoring candidate X is 45 percent plus or minus 6 percent. This means that we can be 95 percent certain that the population percentage is within 6 percent of the sample percentage; in this case, between 39 percent and 51 percent of the population favor candidate X. Or we might say that a 99 percent confidence interval for the percentage of consumers preferring brand Y dog food is 28 percent plus or minus 11 percent, meaning that we can be 99 percent certain that the population percentage is within 11 percent of the sample percentage; in this case, between 17 percent and 39 percent of consumers prefer brand Y.

As with the case of hypothesis testing, however, there is no free lunch. For samples of a given size, the narrower the confidence interval—that is, the more precise the estimate—the less confident we can be of it. Conversely, the wider the confidence interval—that is, the less precise the estimate—the more confident we can be of it. Of course, if we increase the size of the sample, we can both narrow our interval and increase our confidence that it contains the population percentage (or whatever the characteristic or parameter is), but it costs money to increase sample sizes.

Surveys or polls which don't include confidence intervals or margins of error are often misleading. More often than not, surveys do include such confidence intervals, but they don't make it into the news story. Hedging or uncertainty is rarely newsworthy.

If the headline reads that unemployment declined from 7.1 percent to 6.8 percent and doesn't say that the confidence interval is plus or minus 1 percent, one might get the mistaken impression that something good happened. Given the sampling error, however, the "decline" may be nonexistent, or there may even be an increase. If margins of error aren't given, a good rule of thumb is that a random sample of one thousand or more gives an interval sufficiently narrow for most purposes, while a random sample of one hundred or less gives too wide a margin for most purposes.

Many people are surprised at how few individuals pollsters survey to get their results. (The width of the confidence interval for percentages varies inversely as the square root of the size of the sample.) Actually, they generally poll a larger number than is theoretically necessary to compensate for problems associated with getting a random sample. When the random sample selected contains one thousand people, the theoretical 95 percent confidence interval for estimating the percentage who favor candidate X or dog food Y is about plus or minus 3 percent. Pollsters often use plus or minus 4 percent for this sample size because of nonrespondents and other difficulties.

Consider the problems associated with a typical telephone poll. Will the results be affected by leaving out homes without a telephone? What percentage of people refuse to respond, or hang up when they learn a pollster is calling? Since the numbers are chosen at random, what's done when a business phone is reached? What if no one is home, or a child answers the phone? What effect does the sex (or voice or manner) of the telephone interviewer have on the responses? Is the interviewer always careful or even honest in recording the responses? Is the method for choosing exchanges and numbers random? Are the questions leading or biased? Are they comprehensible? Whose answer counts if there are two or more adults at home? What methods are used to weigh the results? If the poll concerns an issue about which opinions are changing rapidly, how are the results affected by spreading the poll out over time?

Similar difficulties apply to personal-interview polls and mail polls as well. Asking leading questions or using an insinuating tone is a common pitfall of personal-interview polls, while an especially important concern in mail polls is avoiding self-selected samples where the most committed, aroused, or otherwise atypical groups of people are more likely to be respondents. (Such self-selected samples sometimes go by the more honest term of "lobby.") The famous 1936 *Literary Digest* poll which predicted Alf Landon would beat Franklin Roosevelt by a three-to-two margin was wrong because only 23 percent of the people who were sent questionnaires returned them, and these generally were wealthier. A similar shortcoming biased the 1948 poll which showed Thomas Dewey beating Harry Truman.

Magazines and newspapers are notorious for announcing biased results based on responses to questionnaires appearing in the periodical. These informal polls rarely come with confidence intervals or any details of the methods used, so the problem of self-selected samples is not always immediately apparent. When feminist author Shere Hite or columnist Ann Landers reports that a surprisingly high percentage of their respondents are having affairs or would rather not have had children, we should automatically ask ourselves who is most likely to answer these questionnaires: someone having an affair or someone reasonably content, someone exasperated by her kids or someone happy with them.

Self-selected samples are not much more informative than a list of correct predictions by a psychic. Unless you get the complete list of predictions or a randomly selected subset, the correct predictions mean nothing. Some of them are bound to turn out true by chance. Similarly, unless your poll sample is randomly selected and not self-selected, the poll results usually mean very little.

In addition to being wise to the problem of self-selected samples, the numerate consumer should also understand the related problem of the self-selected study. If company Y commissions eight studies comparing the relative merits of its product and that of its competitor, and seven of the eight conclude that its competitor's product is superior, it's not hard to predict which study company Y will cite in its television commercials.

As in the chapters on coincidence and pseudoscience, we

see that the desire to filter and emphasize information is at odds with the desire to obtain a random sample. Especially for the innumerate, a few vivid predictions or coincidences often carry more weight than much more conclusive but less striking statistical evidence.

Because of this, it's unclear to me why a collection of intimate profiles or personal stories is so frequently termed a poll. If done well, such a collection is more engaging (even if less convincing) than the typical poll or survey and loses much of its value when wrapped in the ill-fitting shroud of a scientific survey.

Obtaining Personal Information

The name of the game in statistics is the inferring of information about a large population by examining characteristics of a small, randomly selected sample. The techniques involved—from the enumerative induction of Francis Bacon to the theories of hypothesis testing and experimental design of Karl Pearson and R. A. Fisher, the founding fathers of modern statistics—all depend on this (now) obvious insight. Several unusual ways of obtaining information follow.

The first, which will perhaps become increasingly important in an inquisitive age which professes to still value privacy, makes it possible to obtain sensitive information about a group of people without compromising any person's privacy. Assume we have a large group of people and want to discover what percentage of them have engaged in a certain sex act, in order to determine what practices are most likely to lead to AIDS.

What can we do? We ask everyone to take a coin from his or her purse or wallet and direct them to flip it once. Without letting anyone else see the outcome, they should note whether it lands on heads or tails. If the coin lands heads, the person should answer the question honestly: Has he or she ever engaged in the given sexual practice—yes or no? If it comes up tails, the person should simply answer yes. Thus, a yes response could mean one of two things, one quite innocuous (the coin's landing tails), the other potentially embarrassing (engaging in the sex act). Since the experimenter can't know what yes means, people presumably will be honest.

Let's say that 620 of 1,000 responses are yes. What does this indicate about the percentage of people who engage in the sex act? Approximately 500 of the 1,000 people will answer yes simply because the coin landed tails. That leaves 120 people who answered yes out of the 500 who replied to the question honestly (those whose coins landed heads). Thus, 24 percent (120/500) is the estimate for the percentage of people who engage in the sex act.

There are many refinements of this method that can be used to learn more detail, such as how many times people engaged in the sex act. Some variations of the method can be more informally implemented, and could be used by a spy agency to estimate the number of dissidents in an area, or by an advertising agency to estimate the market for a product whose attractiveness people are likely to deny. The raw data for the calculations can come from public sources and, appropriately massaged, can yield surprising conclusions.

Another somewhat uncommon way of obtaining information is the so-called capture-recapture method. Assume we want to know how many fish are in a certain lake. We capture one hundred of them, mark them, and then let them go. After allowing them to disperse about the lake, we catch another hundred fish and see what fraction of them are marked.

If eight of the hundred we capture are marked, then a reasonable estimate of the fraction of marked fish in the whole lake is 8 percent. Since this 8 percent is constituted by the one hundred fish we originally marked, the number of fish in the whole lake can be determined by solving the proportion: 8 (marked fish in the second sampling) is to 100 (the total number of the second sampling) as 100 (the total number marked) is to N (the total number in the lake). N is about 1,250.

Of course, care must be taken that the marked fish don't die as a result of the marking, that they're more or less uniformly distributed about the lake, that the marked ones aren't only the slower or more gullible among the fish, etc. As a way to get a rough estimate, however, the capture-recapture method is effective, and of more generality than the fish example might suggest.

Statistical analyses of works whose authorship is disputed (books of the Bible, *The Federalist Papers*, etc.) also depend on related clever ways of gleaning information from uncooperative (because dead) sources.

Two Theoretical Results

A large part of the attraction of probability theory is the immediacy and intuitive appeal of its practical problems and of the simple principles which enable us to solve many of them. Still, the following two theoretical results are of such fundamental importance that I'd be derelict were I not to mention them at all.

The first is the law of large numbers, one of the most significant though often misunderstood theorems in probability theory, and one which people sometimes invoke to justify all sorts of bizarre conclusions. It states simply that in the long run the difference between the probability of some event and the relative frequency with which it occurs approaches zero.

In the special case of a fair coin, the law of large numbers, first described by James Bernoulli in 1713, tells us that the difference between ½ and the quotient of the total number of heads obtained divided by the total number of flips can be proved to get arbitrarily close to zero as the number of flips increases. Remember from the discussion on losers and fair coins in Chapter 2 that this doesn't mean that the difference between the total number of heads and the total number of tails will get smaller as the number of flips increases; generally, quite the opposite happens. Fair coins behave well in a ratio sense, but not in an absolute sense. And contrary to countless barroom conversations, the law of large numbers doesn't imply the gambler's fallacy: that a head is more likely after a string of tails.

Among other things the law justifies is the experimenter's belief that the average of a bunch of measurements of some quantity should approach the true value of the quantity as the number of measurements increases. It also provides the rationale for the common-sense observation that if a die is rolled N times, the chances that the number of 5s obtained differs much from N/6 gets smaller and smaller as N gets larger.

Succinctly: The law of large numbers gives a theoretical basis for the natural idea that a theoretical probability is some kind of guide to the real world, to what actually happens.

The normal bell-shaped curve seems to describe many phenomena in nature. Why? Another very important theoretical result in probability theory, the so-called central limit theorem, provides the theoretical explanation for the prevalence of this normal Gaussian distribution (named after Carl Friedrich Gauss, one of the greatest mathematicians of the nineteenth or any other century). The central limit theorem states that the sum (or the average) of a large bunch of measurements follows a normal curve even if the individual measurements themselves do not. What does this mean?

Imagine a factory which produces small batteries for toys, and assume that the factory is run by a sadistic engineer who ensures that about 30 percent of the batteries burn out after only five minutes, and the remaining 70 percent last for approximately a thousand hours before burning out. The distribution of the lifetimes of these batteries is clearly not described by a normal bell-shaped curve, but rather by a U-shaped curve consisting of two spikes, one at five minutes and a bigger one at a thousand hours.

Assume now that these batteries come off the assembly line in random order and are packed in boxes of thirty-six. If we decide to determine the average lifetime of the batteries in a box, we'll find it to be about 700 or so; say, 709. If we determine the average lifetime of the batteries in another box of thirty-six, we'll again find the average lifetime to be about 700 or so, perhaps 687. In fact, if we examine many such boxes, the average of the averages will be very close to 700, and what's more fascinating, the distribution of these averages will be approximately normal (bell-shaped), with the right percentage of packages having averages between 680 and 700, or between 700 and 720, and so on.

The central limit theorem states that under a wide variety of circumstances this will always be the case—averages and sums of nonnormally distributed quantities will nevertheless themselves have a normal distribution.

The normal distribution also arises in the measuring process. Here the theorem provides the theoretical support for the fact that the measurements of any quantity tend to follow a normal bell-shaped "error curve" centered on the true value of the quantity being measured. Other quantities which tend to follow a normal distribution might include age-specific heights and weights, water consumption in a city for any given day, widths of machined parts, I.Q.s (whatever it is that they measure), the number of admissions to a large hospital on any given day, distances of darts from a bull's-eye, leaf sizes, breast sizes, or the amount of soda dispensed by a vending machine. All these quantities can be thought of as the average or sum of many factors (genetic, physical, or social), and thus the central limit theorem explains their normal distribution.

Succinctly: Averages (or sums) of quantities tend to follow a normal distribution even when the quantities of which they're the average (or sum) don't.

Correlation and Causation

Correlation and causation are two quite different words, and the innumerate are more prone to mistake them than most. Quite often, two quantities are correlated without either one being the cause of the other.

One common way in which this can occur is for changes in both quantities to be the result of a third factor. A well-known example involves the moderate correlation between milk consumption and the incidence of cancer in various societies. The correlation is probably explained by the relative wealth of these societies, bringing about both increased milk consumption and more cancer due to greater longevity. In fact, any health practice, such as milk drinking, which correlates positively with longevity will probably do the same with cancer incidence.

There is a small negative correlation between death rates per thousand people in various regions of the country and divorce rates per thousand marriages in the same regions. More divorce, less death. Again, a third factor, the age distribution of the various regions, points toward an explanation. Older married couples are less likely to divorce and more likely to die than younger married couples. In fact, because divorce is such a wrenching, stressful experience, it probably raises one's risk of death, and thus the reality is quite contrary to the above misleading correlation. Another example of a correlation mistaken for a cause: In the New Hebrides Islands, body lice were considered a cause of good health. As in many folk observations, there was some evidence for this. When people took sick, their temperatures rose and caused the body lice to seek more hospitable abodes. The lice and good health both departed because of the fever. Similarly, the correlation between the quality of a state's day-care programs and the reported rate of child sex abuse in them is certainly not causal, but merely indicates that better supervision results in more diligent reporting of the incidents which do occur.

Sometimes correlated quantities are causally related, but other "confounding" factors complicate and obscure the causal relations. A negative correlation—for example, between the degree held by a person (B.S., M.A. or M.B.A., Ph.D.) and that person's starting salary—may be clarified once the confounding factor of different types of employers is taken into account. Ph.D.s are more likely to accept relatively lower-paying academic employment than people with bachelor's or master's degrees who go into industry, and thus the higher degree and this latter fact bring about the lower starting salary; a higher degree by itself doesn't lower one's salary. Smoking is without doubt a significant contributory cause of cancer, lung and heart disease, but there are confounding factors having to do with life-style and environment which partially obscured this fact for some years.

There is a small correlation between a woman's being single and her having gone to college. There are many confounding factors, however, and whether there's any causal relation between the two phenomena is unclear, as is its direction, if there is one. It may be that a woman's tendency toward "spinsterhood" is a contributory cause to her attending college, rather than the other way around. Incidentally, *Newsweek* once stated that the chances of a college-educated single woman over thirty-five getting married were smaller than her chances of being killed by a terrorist. The remark was probably intentional hyperbole, but I heard it quoted as fact by a number of media people. If there were an innumeracy-of-the-year award, this statement would be a strong contender for it.

Finally, there are many purely accidental correlations. Studies reporting small nonzero correlations are often merely reporting chance fluctuations, and are about as meaningful as a coin being flipped fifty times and not coming up heads half the time. Too much research in the social sciences, in fact, is a mindless collection of such meaningless data. If property X (say, humor) is defined in this way (number of laughs elicited by a collection of jokes), and property Y (say, self-esteem) is defined in that way (number of yes responses to some list of positive traits), then the correlation coefficient between humor and self-esteem is .217. Worthless stuff.

Regression analysis, which attempts to relate the values of quantity X to those of quantity Y, is a very important tool in statistics but is frequently misused. Too often, we get results similar to the above examples or something like $Y = 2.3 X +$

R, where R is a random quantity whose variability is so large as to inundate the presumed relationship between X and Y.

Such faulty studies are frequently the basis for psychological tests for employment, insurance rates, and creditworthiness. You may make a fine employee or deserve low premiums or a good credit rating, but if your correlatives are perceived to be lacking in some way, you'll have difficulty, too.

Breast Cancer, Muggings, and Wages: Simple Statistical Mistakes

Hypothesis testing and estimates of confidence, regression analysis, and correlation—though all are liable to misinterpretation, the most common sorts of statistical solecisms involve nothing more complicated than fractions and percentages. This section contains a few typical illustrations.

That one out of eleven women will develop breast cancer is a much cited statistic. The figure is misleading, however, in that it applies only to an imaginary sample of women all of whom live to age eighty-five and whose incidence of contracting breast cancer at any given age is the present incidence rate for that age. Only a minority of women live to age eighty-five, and incidence rates are changing and are much higher for older women.

At age forty, approximately one woman in a thousand develops breast cancer each year, whereas at age sixty the rate has risen to one in five hundred. The typical forty-year-old has about a 1.4 percent chance of developing the disease before age fifty and a 3.3 percent chance of developing it before sixty. To exaggerate a bit, the one-in-eleven figure is a little like saying that nine out of ten people will develop age spots, which doesn't mean it should be a major preoccupation of thirty-year-olds.

Another example of a technically correct yet misleading statistic is the fact that heart disease and cancer are the two leading killers of Americans. This is undoubtedly true, but according to the Centers for Disease Control, accidental deaths—

in car accidents, poisonings, drownings, falls, fires, and gun mishaps—result in more lost years of potential life, since the average age of these victims is considerably lower than that of the victims of cancer and heart disease.

The elementary-school topic of percentages is continually being misapplied. Despite a good deal of opinion to the contrary, an item whose price has been increased by 50 percent and then reduced by 50 percent has had a net reduction in price of 25 percent. A dress whose price has been "slashed" 40 percent and then another 40 percent has been reduced in price by 64 percent, not 80 percent.

The new toothpaste which reduces cavities by 200 percent is presumably capable of removing all of one's cavities twice over, maybe once by filling them and once again by placing little bumps on the teeth where they used to be. The 200 percent figure, if it means anything at all, might indicate that the new toothpaste reduces cavities by, say 30 percent, compared to some standard toothpaste's reduction of cavities by 10 percent (the 30 percent reduction being a 200 percent increase of the 10 percent reduction). The latter phrasing, while less misleading, is also less impressive, which explains why it isn't used.

The simple expedient of always asking oneself: "Percentage of what?" is a good one to adopt. If profits are 12 percent, for example, is this 12 percent of costs, of sales, of last year's profits, or of what?

Fractions are another source of frustration for many innumerates. A Presidential candidate in 1980 was reported to have asked his press entourage how to convert $2/7$ to a percentage, explaining that the homework problem was assigned to his son. Whether this report is accurate or not, I'm convinced that a sizable minority of adult Americans wouldn't be able to pass a simple test on percentages, decimals, fractions, and conversions from one to another. Sometimes when I hear that something or other is selling at a fraction of its normal cost, I comment that the fraction is probably $4/3$, and am met with a blank stare.

A man is downtown, he's mugged, and he claims the mugger was a black man. However, when the scene is reenacted many times under comparable lighting conditions by a court investigating the case, the victim correctly identifies the race of the assailant only about 80 percent of the time. What is the probability his assailant was indeed black?

Many people will of course say that the probability is 80 percent, but the correct answer, given certain reasonable assumptions, is considerably lower. Our assumptions are that approximately 90 percent of the population is white and only 10 percent black, that the downtown area in question typifies this racial composition, that neither race is more likely to mug people, and that the victim is equally likely to make misidentifications in both directions, black for white and white for black. Given these premises, in a hundred muggings occurring under similar circumstances, the victim will on average identify twenty-six of the muggers as black—80 percent of the ten who actually were black, or eight, plus 20 percent of the ninety who were white, or eighteen, for a total of twenty-six. Thus, since only eight of the twenty-six identified as black were black, the probability that the victim actually was mugged by a black given that he said he was is only 8/26, or approximately 31 percent!

The calculation is similar to the one on false positive results in drug testing, and, like it, demonstrates that misinterpreting fractions can be a matter of life and death.

According to government figures released in 1980, women earn 59 percent of what men do. Though it's been quoted widely since then, the statistic isn't strong enough to support the burden placed on it. Without further detailed data not included in the study, it's not clear what conclusions are warranted. Does the figure mean that for exactly the same jobs that men perform, women earn 59 percent of the men's salaries? Does the figure take into account the increasing number of women in the work force, and their age and experience? Does it take into account the relatively low-paying jobs many women have (clerical, teaching, nursing, etc.)? Does it take into account the fact that the husband's job generally determines where a married couple will live? Does it take into account the higher percentage of women working toward a short-term goal? The answer to all these questions is no. The bald figure released merely stated that the median income of full-time women workers was 59 percent of that for men.

The purpose of the above questions is not to deny the existence of sexism, which is certainly real enough, but to point out an important example of a statistic which by itself is not very informative. Still, it's always cited and has become what statistician Darrell Huff has called a semi-attached figure, a number taken out of context with little or no information about how it was arrived at or what exactly it means.

When statistics are presented so nakedly, without any information on sample size and composition, methodological protocols and definitions, confidence intervals, significance levels, etc., about all we can do is shrug or, if sufficiently intrigued, try to determine the context on our own. Another sort of statistic often presented nakedly takes this form: the top X percent of the country owns Y percent of its wealth, where X is shockingly small and Y is shockingly big. Most statistics of this type are shockingly misleading, although, once again, I don't mean to deny that there are a lot of economic inequities in this country. The assets that rich individuals and families own are seldom liquid, nor are they of purely personal significance or value. The accounting procedures used to measure these assets are frequently quite artificial, and there are other complicating factors which are obvious with a little thought.

Whether public or private, accounting is a peculiar blend of facts and arbitrary procedures which usually require decoding. Government employment figures jumped significantly in 1983, reflecting nothing more than a decision to count the military among the employed. Similarly, heterosexual AIDS cases rose dramatically when the Haitian category was absorbed into the heterosexual category.

Adding, though pleasant and easy, is often inappropriate. If each of the ten items needed for the manufacture of something or other has risen 8 percent, the total price has risen just 8 percent, not 80 percent. As I mentioned, a misguided local weathercaster once reported that there was a 50 percent chance of rain on Saturday and a 50 percent chance on Sunday, and so, he concluded, "it looks like a 100 percent chance of rain this weekend." Another weathercaster announced that it was going to be twice as warm the next day, since the temperature would rise from 25 to 50.

There's an amusing children's proof that they don't have time for school. One-third of the time they're sleeping, for a total of approximately 122 days. One-eighth of the time they're eating (three hours a day), for a total of about 45 days. One-fourth of the time, or 91 days, is taken up by summer and other vacations, and two-sevenths of the year, 104 days, is weekend. The sum is approximately a year, so they have no time for school.

Such inappropriate additions, although generally not as obvious as that, occur all the time. When determining the total cost of a labor strike or the annual bill for pet care, for example, there's a tendency to add in everything one can think of, even if this results in counting some things several times under different headings, or in neglecting to take account of certain resultant savings. If you believe all such figures, you probably believe that children have no time to attend school.

If you want to impress people, innumerates in particular, with the gravity of a situation, you can always employ the strategy of quoting the absolute number rather than the probability of some rare phenomenon whose underlying base population is large. Doing so is sometimes termed the "broad base" fallacy, and we've already cited a couple of instances of it. Which figure to stress, the number or the probability, depends on the context, but it's useful to be able to translate quickly from one to the other so as not to be overwhelmed by headlines such as "Holiday Carnage Kills 500 Over Four-Day Weekend" (this is about the number killed in any four-day period).

Another example involves the spate of articles a few years ago about the purported link between teenage suicide and the game of "Dungeons and Dragons." The idea was that teenagers became obsessed with the game, somehow lost contact with reality, and ended up killing themselves. The evidence cited was that twenty-eight teenagers who often played the game had committed suicide.

This seems a fairly arresting statistic until two more facts are taken into account. First, the game sold millions of copies, and there are estimates that up to 3 million teenagers played it. Second, in that age group the annual suicide rate is approximately 12 per 100,000. These two facts together suggest that the number of teenage "Dungeons and Dragons" players who could be expected to commit suicide is about 360 (12 × 30)! I don't mean to deny that the game was a causal factor in some of these suicides, but merely to put the matter in perspective.

Odds and Addenda

In this section are several addenda to earlier material in this chapter.

The urge to average can be seductive. Recall the chestnut about the man who reports that, though his head is in the oven and his feet in the refrigerator, he's pretty comfortable on average. Or consider a collection of toy blocks which vary between one and five inches on a side. The average toy block in this collection, we might assume, is three inches on a side. The volume of these same toy blocks varies between 1 and 125 cubic inches. Thus, we might also assume that the average toy block has a volume of 63 cubic inches [(1 + 125)/2 = 63]. Putting the two assumptions together, we conclude that the average toy block in this collection has the interesting property of having three inches to a side and a volume of sixty-three cubic inches!

Sometimes a reliance on averages can have more serious consequences than misshapen cubes. The doctor informs you that you have a dread disease, the average victim of which lives for five years. If this is all you know, there may be some reason for hope. Perhaps two-thirds of the people who suffer from this disease die within a year of its onset, and you've already survived for a couple of years. Maybe the "lucky" one-third of its victims live from ten to forty years. The point is that, if you know only the average survival time and nothing about the distribution of the survival times, it's difficult to plan intelligently.

A numerical example: The fact that the average value of some quantity is 100 might mean that all values of this quantity are between 95 and 105, or that half of them are around 50 and half around 150, or that a fourth of them are 0, half of them are near 50, and a fourth of them are approximately 300, or any number of other distributions which have the same average.

Most quantities do not have nice bell-shaped distribution curves, and the average or mean value of these quantities is of limited importance without some measure of the variability of the distribution and an appreciation of the rough shape of the distribution curve. There are any number of everyday situations in which people develop a good intuition for the distribution curves in question. Fast-food restaurants, for example, provide a product whose average quality is moderate at best but whose variability is very low (aside from speed of service, their most attractive feature). Traditional restaurants generally provide a product of higher average quality but of much greater variability, especially on the downward side.

Someone offers you a choice of two envelopes and tells you one has twice as much money in it as the other. You pick envelope A, open it, and find $100. Envelope B must, thus, have either $200 or $50. When the proposer permits you to change your mind, you figure you have $100 to gain and only $50 to lose by switching your choice, so you take envelope B instead. The question is: Why didn't you choose B in the first place? It's clear that no matter what amount of money was in the envelope originally chosen, given permission to change your mind, you would always do so and take the other envelope. Without any knowledge of the probability of there being various amounts of money in the envelopes, there is no way out of this impasse. Variations of it account for some of the "grass is always greener" mentality that frequently accompanies the release of statistics on income.

One more game. Flip a coin continuously until a tail appears for the first time. If this doesn't happen until the twentieth (or later) flip, you win $1 billion. If the first tail occurs before the twentieth flip, you must pay $100. Would you play?

There's one chance in 524,288 (2^{19}) that you'll win the billion dollars and 524,287 chances in 524,288 that you'll lose $100. Even though you're almost certain to lose any particular bet, when you win (which the law of large numbers predicts will happen about once every 524,288 times on the average), your winnings will more than make up for all your losses. Specifically, your expected or average winnings when playing this game are $(1/524,288) \times (+ \text{ one billion}) + (524,287/524,288) \times (- \text{ one hundred})$, or about $1,800 per bet. Still, most people will choose not to play this game (a variant of the so-called St. Petersburg paradox) despite its average payoff of almost $2,000.

What if you could play as often and as long as you pleased and didn't have to settle up until you were finished playing? Would you play then?

Obtaining random samples is a difficult art, and the pollster doesn't always succeed. Neither, for that matter, does the government. The 1970 draft lottery, wherein numbers from 1 to 366 were placed in little capsules to determine who was to be drafted, was almost certainly unfair. The 31 capsules for January birth dates were placed in a large bin, then the 29 February capsules, and so on up to December's 31 capsules. There was some mixing of the bin along the way, but apparently not enough, since the December dates were disproportionately represented among the early choices, whereas dates from the first months of the year came up near the end significantly more often than chance would dictate. The 1971 lottery used computer-generated tables of random numbers.

Randomness is not easy to obtain when playing cards, either, since shuffling a deck two or three times is not enough to destroy whatever order it might have. As statistician Persi Diaconis has shown, six to eight riffle shuffles are usually necessary. If a deck of cards in a known order is shuffled only two or three times and a card is removed from it and replaced somewhere else in the deck, a good magician can almost always locate it. Using a computer to generate a random ordering to the cards is the best though perhaps an impractical way to get random decks.

One amusing way in which illegal gambling operations obtain daily random numbers which are publicly accessible is to take the hundredth digit, the last and most volatile digit of each day's Dow Jones Industrials, Transportation, and Utilities indices respectively, and juxtapose them. For example, if the Industrials closed at 2213.27, the Transportation stocks at 778.31, and the Utilities at 251.32, then the number for the day would be 712. Since the volatility of these last digits makes them essentially random, every number from 000 to 999 is equally likely to come up. And no one need fear that the numbers are being cooked either, since they appear in the prestigious *Wall Street Journal*, as well as in more plebeian papers.

Randomness is essential, however, not just to ensure fair gambling, polling, and hypothesis testing, but also for modeling any situation which has a large probabilistic component, and for this purpose millions of random numbers are required. How long will someone have to wait in line at a supermarket under different conditions? Design an appropriate program which models the supermarket situation with its various constraints and instruct the computer to run through the program a few million times to see how often different outcomes result. Many mathematical problems are so intractable, and experiments involving them so expensive, that this kind of probabilistic simulation is the only alternative to giving up on them. Even when a problem is easier and it's possible to solve it completely, simulation is frequently faster and cheaper.

For most cases, the pseudorandom numbers generated by computers are good enough. Random for most practical purposes, they are actually generated by a deterministic formula which imposes sufficient order on the numbers as to render them useless for some applications. One such application is to coding theory, which allows government officials, bankers, and others to pass classified sensitive information without fear that it will be unscrambled. In these cases, one mixes pseudorandom numbers from several computers, then incorporates the physical indeterminacy of randomly fluctuating voltages from a "white noise" source.

Slowly emerging is the strange notion that randomness has an economic value.

Statistical significance and practical significance are two different things. A result is statistically significant if it's sufficiently unlikely to have occurred by chance. This by itself doesn't mean much. Several years ago, a study was conducted in which one group of volunteers received a placebo and another group received very large doses of Vitamin C. The group receiving Vitamin C contracted colds at a slightly lower rate than did the control group. The size of the sample was big enough that it was quite unlikely that this effect was due to chance, but the difference in rate was not all that impressive or significant in a practical sense.

A good number of medications have the property that they're demonstrably better than nothing, but not by much. Medicine X which immediately alleviates 3 percent of all headaches in test after test is certainly better than nothing, but how much would you spend on it? You can be sure that it would be advertised as providing relief in a "significant" percentage of cases, but the significance is only statistical.

Usually we encounter the opposite situation: the result is
of potential practical significance but of almost no statistical significance. If some celebrity endorses a brand of dog food, or
some cabdriver disapproves of the mayor's handling of a dilemma, there's obviously no reason to accord statistical significance to these personal expressions. The same is true of women's
magazine quizzes: How to tell if he loves someone else; Does
your man suffer from the Boethius complex?; Which of the seven
types of lover is your man? There's almost never any statistical
validation to the scoring of these quizzes: Why does a score of
62 indicate a man is unfaithful? Maybe he's just getting over his
Boethius complex. Where did this seven-part typology come
from? Though men's magazines often suffer from worse sorts of
idiocies having to do with violence and assassins for hire, they
rarely have these fatuous quizzes in them.

There's a strong human tendency to want everything, and
to deny that trade-offs are usually necessary. Because of their
positions, politicians are often more tempted than most to indulge in this magical thinking. Trade-offs between quality and
price, between speed and thoroughness, between approving a
possibly bad drug and rejecting a possibly good one, between
liberty and equality, etc., are frequently muddled and covered
with a misty gauze, and this decline in clarity is usually an added
cost for everyone.

For example, when the recent decisions by a number of
states to raise the speed limit on certain highways to 65 m.p.h.
and not to impose stiffer penalties on drunk driving were chal-
lenged by safety groups, they were defended with the patently
false assertion that there would be no increase in accident rates,
instead of with a frank acknowledgment of economic and political
factors which outweighed the likely extra deaths. Dozens of
other incidents, many involving the environment and toxic
wastes (money vs. lives), could be cited.

They make a mockery of the usual sentiments about the
pricelessness of every human life. Human lives are priceless in

many ways, but in order to reach reasonable compromises, we must, in effect, place a finite economic value on them. Too often when we do this, however, we make a lot of pious noises to mask how low that value is. I'd prefer less false piety and a considerably higher economic value placed on human lives. Ideally, this value should be infinite, but when it can't be, let's hold the saccharine sentiments. If we're not keenly aware of the choices we're making, we're not likely to work for better ones.

Close

We sail within a vast sphere, ever drifting in uncertainty, driven from end to end.   —Pascal

A man is a small thing, and the night is very large and full of wonders.   —Lord Dunsany

. . .

Probability enters our lives in a number of different ways. The first route is often through randomizing devices such as dice, cards, and roulette wheels. Later we become aware that births, deaths, accidents, economic and even intimate transactions all admit of statistical descriptions. Next we come to realize that any sufficiently complex phenomenon, even if it's completely deterministic, will often be amenable only to probabilistic simulation. Finally, we learn from quantum mechanics that the most fundamental microphysical processes are probabilistic in nature.

Not surprisingly, then, an appreciation for probability takes a long time to develop. In fact, giving due weight to the fortuitous nature of the world is, I think, a mark of maturity and balance. Zealots, true believers, fanatics, and fundamentalists of all types seldom hold any truck with anything as wishy-washy as probability. May they all burn in hell for 10^{10} years (just kidding), or be forced to take a course in probability theory.

In an increasingly complex world full of senseless coincidence, what's required in many situations is not more facts—we're inundated already—but a better command of known facts, and for this a course in probability is invaluable. Statistical tests and confidence intervals, the difference between cause and correlation, conditional probability, independence, and the multiplication principle, the art of estimating and the design of experiments, the notion of expected value and of a probability distribution, as well as the most common examples and counterexamples of all of the above, should be much more widely known. Probability, like logic, is not just for mathematicians anymore. It permeates our lives.

At least part of the motivation for any book is anger, and this book is no exception. I'm distressed by a society which depends so completely on mathematics and science and yet seems so indifferent to the innumeracy and scientific illiteracy of so many of its citizens; with a military that spends more than one quarter of a trillion dollars each year on ever smarter weapons for ever more poorly educated soldiers; and with the media, which invariably become obsessed with this hostage on an airliner, or that baby who has fallen into a well, and seem insufficiently passionate when it comes to addressing problems such as urban crime, environmental deterioration, or poverty.

I'm pained as well at the sham romanticism inherent in the trite phrase "coldly rational" (as if "warmly rational" were some kind of oxymoron); at the rampant silliness of astrology, parapsychology, and other pseudosciences; and at the belief that mathematics is an esoteric discipline with little relation or connection to the "real" world.

Still, irritation with these matters was only part of my incentive. The discrepancies between our pretensions and reality are usually quite extensive, and since number and chance are among our ultimate reality principles, those who possess a keen grasp of these notions may see these discrepancies and incongruities with greater clarity and thus more easily become subject to feelings of absurdity. I think there's something of the divine in these feelings of our absurdity, and they should be cherished, not avoided. They provide perspective on our puny yet exalted position in the world, and are what distinguish us from rats. Anything which permanently dulls us to them is to be opposed, innumeracy included. The desire to arouse a sense of numerical proportion and an appreciation for the irreducibly probabilistic nature of life—this, rather than anger, was the primary motivation for the book.